Developmental Biology

A Guide for Experimental Study

Developmental Biology
A Guide for Experimental Study
Second Edition

Mary S. Tyler
University of Maine

Sinauer Associates • Publishers
Sunderland, Massachusetts 01375

DEVELOPMENTAL BIOLOGY
A Guide For Experimental Study, Second Edition
Copyright © 2000 by Sinauer Associates, Inc. All rights reserved. This book may not be reproduced in whole or in part for any purpose whatever without permission. For information, address:

Sinauer Associates, Inc. Publishers
23 Plumtree Road
Sunderland, Massachusetts 01375-0407 U.S.A.

Fax: 413-549-1118
Email: publish@sinauer.com

Original line drawings by Mary S. Tyler.

Library of Congress Cataloging-in-Publication Data

Tyler, Mary S., 1949–
    Developmental biology: a guide for experimental study / Mary S. Tyler.—2nd ed.
      p. cm.
    Includes bibliographical references.
    ISBN 0-87893-843-5 (pbk.)
    1. Developmental biology—Laboratory manuals. 2. Embryology—Laboratory manuals.
    I. Title.

QH491.T95 2000
571.8'078—dc21

00-026533

Printed in U.S.A.

5 4 3 2 1

For Anna and Matthew

table of Contents

preface
_____

Development is a magnificent and mysterious journey that we all participate in until the cold ground claims us. It is a journey so complex that we travel mostly unaware of the means by which we are speeding along. But the means of travel are well worth our attention. This book invites you to look through the windows that science provides—you may be the one who sees what others have missed.

The Book at a Glance

This guidebook begins with a discussion of the tools you need for studying developing organisms and moves on through the events of development, first in simple and then in more complex organisms.

Chapters 1–3 and 12: Tools of the trade

Much of the study of developmental biology involves craft. The first three chapters and chapter 12 will school you in the methods of this craft, teaching you how to record and report data, how to create your own tools, how to be master of your microscope, and how to prepare tissue for microscopic study.

Chapter 4: A simple organism

Not all organisms use sexual reproduction as the preferred method for passing on their genes. Sex is expensive, burdened as it is with the costs of meiosis. The frugal cellular slime mold, consisting of only a few cell types, seldom indulges in sex, instead undergoing repeated cycles of asexual reproduction. Chapter 4 shows you how to design experiments that use this organism to explore principles of development operating in the realm of simple organization.

Chapter 5: Gametogenesis

In organisms that do use sexual reproduction, gametogenesis marks the beginning of development. Chapter 5 shows you how to sleuth your way through the events that lead to the formation of dimorphic gametes—eggs and sperm, the cells that each carry half a load of genetic material for the new organism.

Chapter 6: Fertilization

Mature gametes have their cellular triggers cocked; fertilization is the explosion that is set off when egg and sperm meet, fusing the two gametes and propelling the new individual off along the road of development. Chapter 6 invites you to explore questions surrounding the mechanisms of this event using sea urchin gametes.

Chapters 6–11: Cleavage, gastrulation, and organogenesis

The journey started at fertilization continues at breakneck speed through cleavage. Cleavage provides the embryo with a large number of cells for continuing the trip. Its product, a beautiful blastula, may be the hollow ball of cells you will see among the sea urchin embryos spiraling across your dish in the experiments in Chapters 6 and 7; a ring of cells surrounding a yolk mass, as you will see in fruit fly embryos in Chapter 8; or a plate of cells balanced on a yolky sphere, as you will see in chick embryos in Chapter 9.

Following cleavage, cells start using a new set of instructions, newly transcribed messenger RNAs (mRNAs), for their developmental journey. Some of these messages instruct the cells to don new clothes, in the form of new cell surface molecules. These molecules identify the cells according to germ layer: ectoderm, mesoderm, or endoderm. Once they put on their new clothes, the cells can't sit still. They begin moving to new positions in the process of gastrulation, which brings the germ layers into positions appropriate for their specific fates. You will see how these movements are choreographed in the sea urchin, fruit fly, and chick in Chapters 6 through 10.

With the germ layers in position, the trip enters a phase of dynamic interactions known as organogenesis. Cells in the embryo relay information back and forth, and these interactions lead to organ differentiation. The embryo is now complete; the embryonic portion of the trip is over. In Chapters 6 through 11 you will glimpse this final leg of the embryonic journey as you raise sea urchin larvae, dissect and experiment with imaginal discs of fruit fly larvae, and test the responses of one of the earliest products of organogenesis in the chick, the embryonic heart.

Chapter 13: Regeneration and pattern formation

Development is a long journey that extends well beyond the event of hatching or birth. One of the fascinating side roads is that of regeneration. Few organisms are capable of such impressive feats of regeneration as those seen in planarians. Forming complete replicas of the intact worm from fragments as small as 1/279th the size of the original organism, the planarian, as you will see in Chapter 13, allows you to venture into the mysteries of regeneration.

Chapters 7 and 14: Development and the environment

Development is always influenced by the environment in which it takes place, and often environmental hazards affecting the delicate embryonic stages of an organism threaten its survival. Organisms such as sea urchins and amphibians, which must entrust their poorly protected embryos to the open wilds, can be particularly vulnerable, and therefore make sensitive barometers of environmental hazards. In Chapter 7, the hazards of ultraviolet radiation, an ever increasing threat, are examined using the delicate sea urchin embryo. In Chapter 14 we end our study of the developmental journey by exploring the woods and streams for evidence of threats to embryos of our cool and thin-skinned amphibian relatives.

One goal of this book is to help students be independent scientists. All the exercises are described in sufficient detail that students should be able to perform them on their own. No secrets have been sequestered to a separate instructor's manual. In addition to instructions and background information, the book provides recipes for solutions, bibliographies for further study, and lists of scientific suppliers.

The illustrations in the book are line drawings designed to clearly delineate components and to draw attention to general principles of structure. Photographs and videos for all chapters can be found on the supplementary CD-ROM, *Vade Mecum*.

Vade Mecum—A Supplementary CD-ROM

Vade Mecum: An Interactive Guide to Developmental Biology, by Mary S. Tyler and Ronald N. Kozlowski, and published by Sinauer Associates, was developed primarily to augment this laboratory manual. The goal of the combined manual and CD is to provide students with all that they need to be truly independent learners. Each chapter of the laboratory manual is represented by a chapter on the CD, including over 130 videos and 300 labeled photographs that explain the development of the organisms and give step-by-step explanations of techniques. For example, where useful, color-coding is superimposed on living embryos to illustrate positioning of different germ layers. A complete set of cross sections of a 33-hour chick embryo and whole mounts with definitions of terms are included. A "virtual microscope" section shows how to achieve Koehler and dark-field illumination and how to use polarizing filters. For each chapter, there are study questions and websites listed. A section on laboratory safety is also included. *Vade Mecum* can be purchased as a stand-alone item, or bundled with the laboratory manual at reduced cost, and also comes packaged with Scott Gilbert's *Developmental Biology,* Sixth Edition.

Acknowledgments

Many people have contributed ideas, inspiration, and hard work to these pages, and to them I am very grateful. I thank all the reviewers who added their wisdom and careful comments:

Nikki Adams, John Dearborn, Brian Hall, Malcolm Hunter, Richard Kessin, John Lucchesi, Robert Mead, David McClay, Drew Noden, John Ringo, Brian Sullivan, Seth Tyler, and Bonnie Wood. To those at Sinauer Associates, publisher Andrew Sinauer, production specialist Chris Small, and editors Kathaleen Emerson, Chelsea Holabird, and Carol Wigg, I extend a special warm thanks for bringing the book in its second edition to reality. Their creativity, attention to detail, and generous patience and support were invaluable. To Ryan Genz I am indebted for his creative cover design. And to my always-curious students, I am beholden for their attentiveness, which has taught me to see more than I would alone.

Mary S. Tyler
Orono, Maine

chapter 1 Getting Started

As you embark on your exploration of developmental biology, the most important thing to bring with you is curiosity. Ask questions, for these will lead to hypotheses, and hypotheses open the mind to discoveries. Keep careful records of your questions and answers. Above all, enjoy exploring.

Things You Will Need

The list of things you will need to bring is short.

Laboratory notebook   A standard 8.5"×11" looseleaf notebook is the most useful. The pages are large, and you can add pages where you need them.

Drawing paper   Any thickness, white, unlined, 8.5"×11" paper. Most experiments will require drawings to record data. If you erase a lot or simply enjoy putting pencil to paper, you might consider having the more expensive heavyweight paper that holds up under many erasures.

Colored pencils   Red, yellow, blue, and green. When studying germ layers, you will be color-coding your diagrams according to the standard embryological code: red = mesoderm, yellow = endoderm, and blue = ectoderm. Green often is used to designate neural crest.

Dissecting kit   Blunt probe, dissecting needle, scalpel, and scissors. For the most part, you can use microdissecting tools that you make yourself. However, a dissecting kit will be useful at times.

Your Laboratory Notebook

Everything pertaining to the laboratory is to be kept in your laboratory notebook: notes, drawings, data, speculations, thoughts in the middle of the night. This is a working notebook of the type every experimenter keeps. It is not necessary for your records to look like a polished, neat report (unless, of course, this is your natural style). Your notebook must be orderly and complete, however. Since the quality of your science can be seen from the quality of your laboratory notebook, your laboratory instructor may check yours from time to time. To help make the quality of your science the very highest, here are some important hints.

Always write in pencil   Laboratory spills can wipe out whole pages of data written in pen.

Always take careful, complete notes on all observations   Minutiae often lead to great discoveries. Record with descriptions, measurements, and drawings wherever possible. Diagram to scale, noting the scale used. Drawings need not be artistic, but they do need to be accurate. If you can see it, you can draw it. If you can't draw it, you're not seeing it. Just look harder.

Never recopy notes   Valuable information can be lost. Remember, this is a working notebook. Follow a specific format so your notes are orderly, and make sure you've written neatly enough so that you can read it. It's your notebook—not the instructor's or anyone else's.

ALT-1
Date, Time

TITLE OF STUDY

General procedures

This section should describe the general experimental procedures used. In many cases, these will be the procedures given in a laboratory manual. You do not need to recopy these, but you do need to indicate any changes or additions you make.

Experimental setup

This section describes the specific procedures for each individual experiment within the study. You number each experiment and state exactly what you did in setting up this experiment. You also state what you are testing: What questions are you asking of the organism?

Observations and results

This section is a careful record of everything you do and observe during the course of the study. Every time you make a set of observations, you start by dating the page, and then recording your observations using the code you invented for each experiment. Observe, measure, and draw. Don't crowd information. Leave space between your notes for one experiment and those for the next. Remember, these are notes. They need not be in complete sentences, but they should be extremely precise and clear. These are your data—the most important part of all your work. Hypotheses and conclusions can be wrong; data must stand the test of time.

Summary of results

When the experiments are wrapped up, the excitement builds. You can now collate your data to figure out what they're telling you. Assemble the data into graphs and charts; look for trends and timing of events; determine how the data from one experiment compare with those from the others. Milk the data. And remember, this section is fact—it is a summarization of everything your organism has told you.

Conclusions

Here the fun explodes to party levels. Where before you were adhering to just-the-facts-Mac, you now can expand into conjecture. Decide what those facts could mean. Speculate, theorize. Use whatever background information you have from courses, textbooks, laboratory manuals, or other literature. This section does not have to pass the test of time—you're allowed to be wrong. Think hard and have fun.

Figure 1.1   A typical format for recording data in your laboratory notebook.

A format for your notebook

Any research scientist establishes a format for recording data. The one shown in Figure 1.1 is typical. You will need to invent a coding system for numbering your studies and each experiment within a study. In a classroom full of scientists, it is convenient to use your initials as part of the code. For example, if your initials are A.L.T., ALT-1 could refer to your first study. ALT-1-1 then could refer to the first experiment in your study, ALT-1-2 to the second, and so on.

Formal Laboratory Reports

A formal laboratory report is no more and no less than a scientific paper in a format for publication. Be impressed, and know that you can produce work worthy of publication. In order to publish, though, one must repeat experiments a number of times, and you probably won't have the luxury of this in the confines of a one-semester course. Your laboratory reports, therefore, will be a training ground for the scientific papers many of you will publish in the future (perhaps sooner than you think). Undergraduate students are perfectly capable of producing good, publishable papers, just as seasoned scientists are perfectly capable of producing bad ones. The secret is to practice the former and be able to recognize the latter.

Presented here is a format for a formal scientific paper. Whether you use the entire format or a truncated form of it, you should at least know the anatomy of a scientific publication. Each section has a specific purpose. If you understand that purpose and adhere to it, your chances of producing excellent papers will increase.

Anatomy of a scientific paper

Abstract   This is a concise rendering of the question you posed, the methods you used to investigate it, your results, and your conclusions. Usually only 100–200 words, it is almost always composed after the rest of the paper is written, even though it is placed first.

Introduction   This is a statement of the problem you are investigating and its significance, with enough background information to understand the next two sections of the paper. It is not a long section; most of the background information that needs to be included in the paper can be saved for the Discussion. Your main objective in the introduction is to get the readers to understand what you were asking and why you bothered to ask it. You want, very concisely and effectively, to hit them between the eyes with the meaning of your work, showing them how it fits into the bigger scheme of things. If you don't succeed here, they may not bother to read further.

Materials and methods   This is the boring section. It is a very careful explanation of your methods, the precise materials you used, and where you got those materials. List species names, collecting sites, and suppliers of reagents, along with an explanation of the experimental design for each experiment. This allows you and others to repeat your experiments.

Results   This is a just-the-facts section. It is the section that should remain when all others burn in the fires of future truths and hindsight. It is created from the "Summary of Results" section from your laboratory notebook as described above. Data are presented in digested form, summarized in charts, graphs, and descriptions. You do not conjecture as to what these data are telling you—you simply present them.

Discussion   This is where you can let go and wax eloquent. A discussion is usually divided into two parts: first a statement of your conclusions (the "Conclusions" section from your laboratory notebook), then a discussion of the significance of these conclusions as they relate to other studies published in the literature. You must state how your results corroborate or disagree with previous work. In a paper for publication, you need to have done an exhaustive literature search before writing this section. (For a student laboratory report, your instructor should give you guidelines on the number of papers to be considered in your Discussion.) When you cite another person's work, you need to write in parentheses the names of the authors and date of their publication. Readers then can find the full citation of this work in your Literature Cited section. You want to leave the reader with a full understanding of the importance of your results and their implications as they relate to bigger questions.

Literature cited   This is your bibliography. It is a listing of all the literature you cited throughout your paper. The format for this is rigid, and each journal will have its own rules. Whatever format you use, you must be consistent. An example of one acceptable format is given below. Journal names are abbreviated (usually using the abbreviation the journal itself uses or that used by *Biological Abstracts*, which has a listing called *Serial Sources of the Biosis Data Base* that is published each year). Following the journal name comes the volume number, then the page numbers of the article. Notice that the titles of books and journals, as well as species names and the names of genes, are usually italicized or underlined. You can find more examples of citations in the Selected Bibliography section at the end of each chapter in this book.

Journal article:

Smith, J. S. and J. P. Jones 1999. The significance of nothing. *J. Biol. Nonsense* 14: 13–17.

Article from a book:

Watler, J. N. 1999. *Drosophila* without *per* still uses clock. In *The Anatomy of a Fly,* M. I. Fruit and R. U. Fly (eds.). Unbelievably Esoteric Press, New York, pp. 78–91.

Book:

Tyler, M. S. 2000. *I Didn't Say That.* Ununited Professors Press, Wabash, CT.

Some Hints about Writing

Writing often does not come easily to scientists, and I'm convinced this is primarily the result of beating ourselves out of our natural styles. When you are caught in the grip of unintelligible prose, I suggest you apply these few rules to the afflicted areas. The cure is often immediate.

Don't pull teeth   Writing should never be painful. If you are not having fun, the writing is probably not very good. Relax. Don't be an editor as you write. Edit later. Enjoy being yourself on paper.

Hit, don't be hit   The most common fault that makes writing dense and indecipherable is the use of the passive rather than the active voice. Instead of saying "the egg was fertilized by the sperm," say "the sperm fertilized the egg." In more complex sentences, this invariably clears away clouds of confusion.

Accent the important   All sentences have a natural stress point. Make sure the important information in your sentence is at this stress point, usually at the end of the sentence.

Keep the train cars connected   Sentences of a paragraph, like the cars of a train, must be connected. This is often done by making the subject of one sentence the subject of the next, or by placing old information before new in a sentence. Whatever the technique, one sentence must lead naturally to the next.

Read good writers   You learn by example. Choose writing that is effective and dissect it. Ask, What did the writer do that makes it good?

Read about writing   There are a number of books and articles about writing, some of which are listed in the Selected Bibliography at the end of this chapter. (If you devote no more than an hour to this step, read the 1990 article, "The Science of Scientific Writing" by Gopen and Swan. Its examples of obtuse writing will make it obvious that a great deal of scientific writing is bad writing. You can do better.)

Using the Library

The library is a source of buried treasure, so bring the correct digging tools. Find out what computerized databases are available. Your library's website is a good place to start. Some of the databases that will be of use are Expanded Academic Index, Medline, UnCover, and WorldCat.

Look up websites for specific journals; often they allow you to conduct electronic searches of their articles. Abstracts and sometimes the entire text of articles may be available at these sites. A few examples include *BioEssays* (www.interscience.wiley.com), *Developmental Biology* (www.idealibrary.com), *FASEB Journal* (www.fasebj.org), *Nature* (www.nature.com), *Science* (www.sciencemag.org), *Scientific American* (www.sciam.com), and *Science News* (www.sciencenews.org).

Electronic databases are exceedingly useful, but realize they are only one method used in compiling a reading list on a subject. The most important first article you can get is a recent review article about the subject. Use its bibliography to make your first short list of papers to read.

Also, learn to use scientific indices; some of the useful ones are listed below. *Biological Abstracts* and *Science Citation Index* are two that are available in electronic form over the web—your library may subscribe to them. These indices always include instructions for their use, but they still may be confusing. Ask a librarian to help—librarians consider this part of their job, so don't be shy. Often they have instruction sheets they can give you.

And finally, browse the current literature. The current issues of journals are usually displayed in a separate place from the bound back issues. Know which journals deal with your subject and get used to browsing through them. Included below are lists of specialized developmental biology journals and the more general journals that are important to developmental biologists. (These are only short lists; many excellent, important journals have been left off.) If you scan only ten of these on a weekly basis, you quickly will be learning the art of being a scientist.

Scientific indices
  Biological Abstracts
  Biology Digest

Science Citation Index
Zoological Record

Journals specific to developmental biology
BioEssays
Cell Differentiation
Development (formerly *Journal of Embryology and Experimental Morphology*)
Developmental Biology
Development, Growth, and Differentiation
Differentiation
Roux's Archives of Developmental Biology (alternate title *Wilhelm Roux' Archiv;* formerly *Wilhelm Roux' Archiv fur Entwicklungsmechanik der Organismen*

Journals that cover development and other topics
American Zoologist
Cell
Experimental Cell Research
Journal of Biological Chemistry
Journal of Cell Biology
Journal of Experimental Zoology
Nature
Proceedings of the National Academy of Science (USA)
Science

Journals for the broader audience
American Scientist
Scientific American
Science News

Accompanying Materials to This Manual

The CD-ROM *Vade Mecum: An Interactive Guide to Developmental Biology*, by Mary S. Tyler and Ronald N. Kozlowski, 2000, Sinauer Associates, Sunderland, MA, is designed to accompany this laboratory manual. It gives background information, illustrates techniques, and shows the living organisms with over 130 QuickTime movies and over 300 labeled pictures. Its purpose is to allow students and instructors to work independent of further instruction.

The CD-ROM *FlyCycle-II*, by Mary S. Tyler and Ronald N. Kozlowski, 2000, Sinauer Associates, Sunderland, MA, is an adaptation of the 45-minute film, *Fly Cycle: The Lives of a Fly*, Drosophila melanogaster, 1996, by M. S. Tyler, J. W. Schnetzer and D. Tartaglia, Sinauer Associates, Sunderland, MA. This covers the life cycle of the fruit fly as well as a number of the mutants used in research.*

There are a number of superb texts and films in developmental biology. Listed below are three that are cited, along with text page numbers, in each chapter where relevant to help coordinate lecture and laboratory. If you are using a different text in your course, just use the text's index to find the pages relevant to each laboratory exercise.

---

*An earlier version of the CD-ROM, *FlyCycle* is also available as *FlyCycle* by M. S. Tyler, R. N. Kozlowski, and L. Iten, 1998, Sinauer Associates, Sunderland, MA.

Text: Gilbert, Scott F. 2000. *Developmental Biology*, 6th Ed. Sinauer Associates, Sunderland, MA.

Films: Fink, Rachel (ed.). 1991. *A Dozen Eggs: Time-Lapse Microscopy of Normal Development*. Sinauer Associates, Sunderland, MA.

Fink, Rachel (ed.). 1995. *CELLebrations*. Sinauer Associates, Sunderland, MA.

Tyler, Mary S., Jamie W. Schnetzer and David Tartaglia. 1996. *Fly Cycle, the Lives of a Fly*, Drosophila melanogaster. Sinauer Associates, Sunderland, MA.

Selected Bibliography

Gopen, G. D. and J. A. Swan. 1990. The science of scientific writing. *Am. Sci.* 78: 550–558. This article gives excellent, concise descriptions of many common mistakes that make scientific writing unnecessarily obtuse.

McMillan, V. E. 1988. *Writing Papers in the Biological Sciences*. St. Martin's Press, New York. This is very useful for its suggestions on presentation of data, formats for literature citations, and tips on how to write clearly and accurately.

Meyer, A. W. 1939. *The Rise of Embryology*. Stanford University Press, Stanford, CA. It is instructive to see the laboratory notebooks of famous scientists; unfortunately, facsimiles are not readily available. This book, however, includes a number of plates showing original drawings and text by scientists, including drawings of chick development by Marcello Malpighi and diagrams of mammalian sperm by Antonj van Leeuwenhoek.

Moore, R. 1992. *Writing To Learn Biology*. Saunders College Publishing, Philadelphia. An excellent, delightfully written book. Not only will it help you to write clearly and to recognize bad writing, but it will entertain you along the way with a myriad of humorous and clever examples.

Pechenik, J. A. 1997. *A Short Guide to Writing About Biology*, 3rd Ed. Longman Publ., New York. An excellent writing manual that includes chapters on writing laboratory reports and giving oral presentations.

Slack, J. M. 1983. *From Egg to Embryo: Determinative Events in Early Development*. Cambridge University Press, London. This is a sophisticated, well-written summary of embryonic development. It will introduce you to the major questions rattling around in the minds of present-day developmental biologists.

Strunk, W., and E. B. White 1999. *The Elements of Style*, 4th Ed. Simon and Schuster, Inc., New York. No writing manual has ever topped this one for its helpfulness and conciseness.

Wolpert, L. 1991. *The Triumph of the Embryo*. Oxford University Press, Oxford. Truly a delightful book about embryonic development, this serves as an excellent introduction to the subject and can't fail to entice you further into the field.

2 *Embryological Tools*

Most of an embryologist's tools are so fine that they have to be made by hand. The ones you will be making are those used most routinely by embryologists. The only indispensable microdissection tool that you will not be making is fine forceps. You can usually buy these (#5) from any biological or electron microscopy supply company (see the end of this chapter for addresses), or your laboratory may have a supply from which you can borrow.

Tools to be used on living material must be kept separate from those that have been in fixative. Your dissecting tools from other labs (such as comparative anatomy, where you were up to your elbows in fixed material) must never touch those that you make for living material. Fixative is nearly impossible to remove completely from a tool, and the smallest trace can severely damage living embryological material.

Microknives

Knives as fine as the iridectomy knives used by eye surgeons can be made from chips of razor blades mounted on wooden dowels. ***Protective eye gear and leather work gloves must be worn when breaking the razor blades.*** Break a razor blade into a number of small pieces by cutting it with heavy-duty scissors, aviation tin snips, or a pair of dikes. If there is an exhaust hood or dissecting chamber in the lab, cutting should be done behind the protection of the glass.

The ideal knife is shaped like a half-spear with a long shank (see Figure 2.1A). Prepare a knife handle by first using a pencil sharpener to give a slight taper to the end of a 6-inch length of quarter-inch dowel, then clamp the dowel in a vise and make a slit in the tapered end by lightly hammering a heavy-duty scalpel blade half an inch into the dowel end. Holding a prepared piece of razor blade with forceps, insert the shank of the piece into the slit in the dowel. If the blade needs further securing, glue it in place with fingernail polish. Make two microknives. Store the knives with a length of a soft-drink straw over the end to protect the knife blade. Microknives can be sterilized before use in 70% ethanol or in boiling water.

Author's note: Many of the techniques in toolmaking and sterile technique that I pass on were taught to me by my major professor, Dr. H. Eugene Lehman. The many hours I spent in the quiet of his tiny lab were hours of discovery and wonder. It was there, in those cramped, hot quarters, that I was initiated as an embryologist to the mysteries of my subject and the tools that would help me explore those mysteries. I stand in a long line of oral and bench-tutoring tradition, and it is this that I try to pass on to you in these few pages.

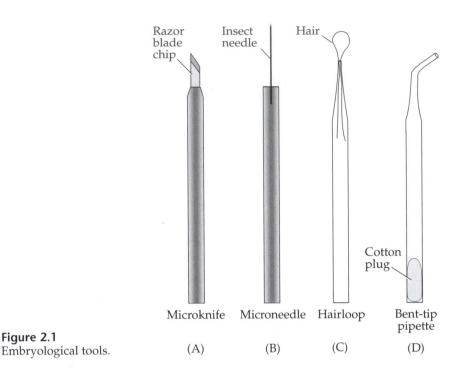

Figure 2.1
Embryological tools.

Microneedles

Microneedles (Figure 2.1B) are extremely valuable for fine manipulations and can be used as microscissors by drawing one needle past another. You will be using insect pins (#00) mounted in the ends of wooden dowels. Break the plastic ball off the end of an insect pin (you can crush it with forceps). If you would like a shorter pin, cut off a length with wire cutters. Hold a 6-inch length of quarter-inch dowel in a vise; with a hammer and fine finishing nail, make a quarter-inch or longer hole in the end of the dowel. Force the blunt end of the insect pin as far down into the dowel as you can, securing it in place with fingernail polish. Make two needles. Store needles with a length of soft-drink straw over the end to protect the tip. Microneedles can be sterilized before use in 70% ethanol or in boiling water.

Hairloops

Hairloops (Figure 2.1C) are one of those romantic additions to an embryologist's arsenal. Dating from the days of Hans Spemann, when experimental embryology was in its infancy, they have been used to gently coax delicate tissues around a culture dish or to deliver grafts to a graft site. Blond baby's hair makes the finest quality loops, and for years embryologists have been known to hang around barber shops waiting for the first haircut of a blond child. You will probably want to experiment with several types of hair. Fine blond hair will make the smallest loops. Coarser hair will make larger loops. Basically, you want to force the hair into the smallest loop possible to give you a firm instrument.

You will be using a Pasteur pipette to hold the hair. Since the opening of the pipette is too large to hold the hair securely, you will be narrowing the opening by heating the bill and drawing it out. ***You must use protective eye gear for these procedures.*** Use heavy-duty forceps to hold the pipette at either end, and pass about

2 inches of the bill of the pipette back and forth through the flame of a Bunsen burner to heat it evenly. When the glass is soft, remove it from the flame and quickly pull on both ends using the heavy forceps. This will stretch the glass into a long fiber. When the glass is cool, use a diamond marking pencil to score the glass at a level where the diameter of the bore is approximately 1 mm. Break the glass at this point by pulling in opposite directions on either side of the score line. This will give a clean, perfect break. Do not bend the glass to break it. This will shatter the glass and give a ragged break.

Take a length of hair (at least several inches; longer lengths tend to be easier to manipulate) and push both ends into the small end of the pipette. Make sure that the protruding loop is no more than 5 mm long. A large, floppy loop is relatively useless. Melt a small amount of paraffin on a glass slide by placing the slide on a hot plate set at a low temperature or by holding the slide with forceps in a flame. Touch the end of the pipette to the edge of the puddle of melted paraffin. Capillary action will draw the paraffin up into the pipette. Allow the paraffin to harden. Any excess paraffin on the hairloop can be removed by warming a slide, placing filter paper on the warm slide, and touching the hair to the filter paper. Make one or more hairloops. Hairloops can be sterilized in 70% ethanol but not in boiling water or in a flame.

Pipettes

Pipettes are among of the most valuable instruments you can have. The appropriate bore size can perform magic at the bench, and you won't know the appropriate size until you're sitting there needing it. It therefore is very important to have made a variety of sizes in advance.

Always wear protective eye gear when breaking glass or using an open flame. To make widemouthed pipettes, break off the end of a Pasteur pipette where the diameter of the bill begins to widen. As you did when making hairloops, score the glass with a diamond marking pencil where you want the break to be, and give a straight pull on the glass on either side of the score line. To avoid a rough opening, fire-polish the end by passing it quickly through a flame. If you heat the end too much, the opening will get smaller or even close. Make several widemouthed pipettes.

To make micropipettes, you will be using Pasteur pipettes and pulling them in a Bunsen burner flame, as you did when making hairloops. Hold both ends of the pipette with heavy-duty forceps and heat approximately 2 inches of the bill until the glass is soft. Quickly remove the pipette from the flame and pull on both ends until the hot glass is pulled out into a thin fiber. When the glass is cool, use a diamond marking pencil to score the glass at a level that will give a bore that is smaller than the original opening. Break the glass by giving a straight pull on either side of the score line. If you choose to fire-polish the end, do so with a very quick swipe through the flame. These small openings occlude easily, and often a clean break is better left unpolished.

You will be putting a 20° bend in the ends of some of your pipettes (Figure 2.1D). This creates a cradle for your tissues when you are transporting them around a dish. The tissue settles into the bend of the pipette rather than getting stuck way up in the bill of the pipette. To create the bend, heat approximately 2 inches of the bill until the glass is soft. Quickly remove the pipette from the flame, and with heavy-duty forceps, make a 20° bend in the glass about one-half inch from the end. Put bends in some or all of your micropipettes as well as in several unmodified short-billed pipettes.

All your pipettes must be plugged with cotton (Figure 2.1D). The cotton will filter any contaminants out of the air when the pipette is in use, and it is absolutely necessary for sterile procedures. Take a small wad of cotton and flatten it into a nickel-sized patty. Then take a much smaller wad—about the size of a thick pencil point—and place it in the center of your nickel patty. Use a blunt probe to push against the smaller wad as you push the entire cotton round into the large open end of the pipette. Do not twist the cotton while pushing it in, because this will create air channels that will serve as avenues for contaminants. Make sure no cotton protrudes from the end of the pipette, since this will get caught in the pipette bulb when using it and will cause the plug to be pulled out. Put cotton plugs in all the pipettes that you made as well as in several unmodified short-billed Pasteur pipettes.

The pipettes should be sterilized in an autoclave. To prepare the pipettes for autoclaving, wrap them in aluminum foil and attach a name tag written in pencil to the bundle using string or masking tape. You may want to first separate your pipettes according to size, making smaller bundles and marking these according to size. If you do, do a second wrapping so that they are together in a single larger bundle that is tagged with your name. Your instructor will probably autoclave the pipettes for you, doing those of the entire class at the same time.

Embryo Spoons

An embryo spoon is one of those tools you seldom use and can't do without. It is nothing more than a perforated spoon that allows you to pick up embryos and eggs in the several-millimeter to several-centimeter range without having them float off the spoon. A metal embryo spoon can be bought for around 75 dollars, or you can make one from a cast-off plastic spoon for less than 2 cents.

A really cheap, thin, relatively small plastic spoon is best. Heat a clean dissecting needle in a Bunsen burner until it is glowing red, then perforate the bowl of the spoon, making 12–15 holes, evenly spaced, and enlarging each hole with the heated needle so that each is approximately 2 mm in diameter. You will have to reheat the needle periodically. Once the holes are made, you will notice that their edges are very rough. Sand both the upper and lower surface of the spoon with sandpaper until your fingers can no longer feel any rough edges. This sanding is extremely important, because a rough edge can seriously damage an embryo.

Test your spoon by putting water in it. The water should drain through the holes. If the spoon holds water instead, the holes need to be enlarged.

Your plastic embryo spoon can be sterilized in 70% ethanol, but not, of course, in either boiling water or a flame.

Instrument Tray

A tray holds your instruments so that they are both protected and ready for use. The tray should be made out of metal, since this is easily sterilized. A piece of sheet metal 8″ × 4″ can be bent so that it looks like a three-sided pencil tray (Figure 2.2). Use metal cutters to cut V-shaped grooves on each side, so that pairs of grooves line up. The instruments will rest in these grooves. Your instrument tray

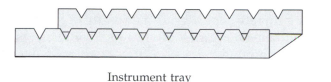

Instrument tray

Figure 2.2
Instrument tray.

can be sterilized in a variety of ways, but wiping it down with 70% alcohol is usually the simplest method.

Retooling Metal Instruments

Tools such as forceps and scalpels must constantly be rehoned to keep them useful. It is best to use a waterstone for this rather than an oilstone, to avoid oil contamination on your tools. An Arkansas stone (though often used with oil, it can also be used with water) is an economical, high-quality stone. Japanese waterstones are also excellent, but are usually more expensive. None are cheap. Use a soft Arkansas stone or an 800–1200 grit Japanese waterstone for the first coarse honing, and a hard Arkansas stone or a 4000–8000 grit Japanese waterstone for the final polish. Always keep the stone wet with water while sharpening—this keeps metal particles from clogging the pores of the stone—and use the whole surface of the stone to avoid uneven wear. Wash the stone with water after use to remove metal residues and store the stone dry in a soft cloth bag.

For a scalpel, as with knives such as those in your kitchen or pocket, sharpening should be done by stroking the blade against the stone in a forward stroke, with the blade tipped at a shallow angle and the cutting edge pointing forward. When the tilt is correct, you will feel the blade biting slightly into the stone. Notice that you are pushing the cutting edge of the blade into the stone, rather than away from it. Trust me, this is the right way to do it. After finishing one stroke, turn the blade over and repeat the stroke on the other side. Check the blade under a dissecting scope. Continue the coarse honing until there are no obvious nicks. Always follow a coarse hone with a fine hone, which should take few strokes in comparison. Always clean the blade carefully between a coarse and fine hone, and after the final hone. Clean the knife by using water and a soft cloth or paper, and wipe toward the cutting edge, never away from it.

The retooling of forceps is an art that is easily learned but seldom taught. (This method will not work on your heavy stainless steel forceps, but will on the very fine—#5 or less—forceps, whose tips are so easily damaged.) First examine your forceps under the dissecting scope, closing them slowly to determine if the tips meet precisely, or if they are out of alignment. If the tips are badly damaged, hold the forceps straight up and down and rub them vigorously against the coarse stone to remove the damaged tip. Then, holding the forceps closed, at a slight angle against the stone, and oriented so that the two halves are both touching the stone with equal pressure, grind back and forth. Turn the forceps over and repeat the operation. Now grind the two remaining sides. Examine the forceps under the dissecting scope. You should be regaining a fine tip to the forceps, and the two halves should meet precisely. Do the final set of grindings on the fine stone.

If the tips are not damaged but the two prongs are out of alignment, then grind the sides as described above.

If your forceps are merely bent slightly at their tip, this sometimes can be fixed by putting the stone between the two prongs and pulling the forceps backward, putting equal force on the two halves. With enough pressure, the ends usually straighten as they come off the stone.

Sterile Technique

Sterile technique is no more than a matter of common sense. If you simply imagine that everything about you is filthy—your hands, your hair, your breath—and that germs are raining down from the heavens, and let this principle guide your actions, then you will master sterile technique quickly.

The ammunition that you can use against this filth is limited: it must be germicidal but leave no residue that could harm your tissues. Our list includes ultraviolet (UV) light, 70% alcohol (any higher or lower percentage is less germicidal), wet heat (such as boiling water or an autoclave), and dry heat (such as an oven, autoclave, or open flame). All work areas, instruments, solutions, and cultureware must be sterilized.

The workbench

A sterile workplace is extremely important. If you have the luxury of building your own workplace, you will want to construct a culture hood (Figure 2.3), paint it with a high-gloss enamel, and install two lights, one a work light and the other a germicidal UV light. The UV light is extremely hazardous to germs, to living things in general, and to your eyes. When the UV light is on, the culture hood must be covered with an opaque curtain (coated bench paper works well) to avoid eye exposure. An hour of exposure to UV is sufficient to sterilize the area, but researchers often leave the UV light on overnight before they work. *Never have the UV light on when you are working under the hood, and never leave your cultures in the hood with the UV light on.*

Most of you will not have the opportunity to build your own culture hood. In this case, your work space on the lab table should be wiped down with 70% alcohol. You must also use 70% alcohol to wipe down the stage of your dissecting scope, if you are using one, as well as its knobs, your instrument tray, and any other items that you will be using.

Dissecting tools

Your dissecting tools must be sterilized, and then resterilized after each use. Most tools can be sterilized by soaking them for 10 minutes in 70% alcohol, by placing

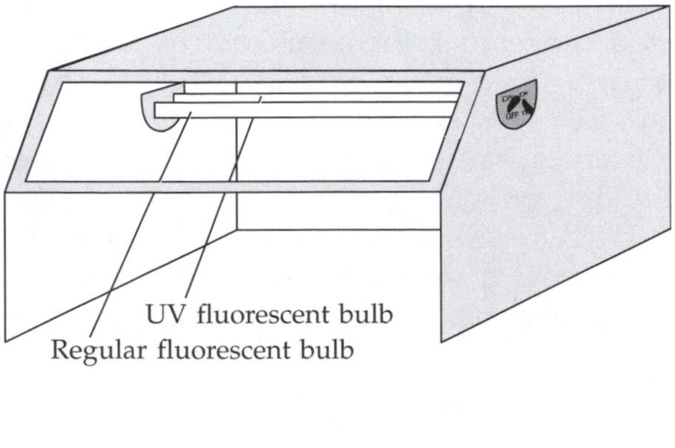

Figure 2.3
Homemade culture hood.

UV fluorescent bulb
Regular fluorescent bulb

them in boiling water, or by autoclaving them. Specific suggestions for each tool were given in the earlier toolmaking section. For most of your tools, it will suffice to soak them in 70% alcohol. Obtain a lidded jar suitable for this; push cotton to the bottom as a cushion for the tips of your instruments. Label the jar in pencil with your name and fill it with 70% alcohol. (The same alcohol can be used many times.) When you are doing an experiment, sterilize your instruments, tips down, in this alcohol, then rest them on an instrument tray to dry. No alcohol must be left on the instrument when it comes in contact with your culture tissue. Alcohol kills. Some researchers prefer to sandwich their instruments between the folds of a paper towel that has previously been soaked in 70% alcohol instead of using an open instrument tray. This is an extremely effective method of keeping germs away from the instruments, but it has the disadvantage of making the researcher fish blindly for the instrument that is needed. Figure out a method that best suits you. Every time you use an instrument, dip it back in the alcohol (a dip will do), and return it to the instrument tray or sterile towel envelope.

Glassware

Any glassware that is used must have been cleaned carefully using detergent such as Alconox that is safe for culture use. Glassware must be rinsed very carefully, then re-rinsed in distilled water and air-dried. It is then wrapped in aluminum foil, folding over each seam twice so that the seams are airtight. Glassware with no cotton stoppers or plastic tops can be autoclaved in dry heat or sterilized in an oven at 400°F for an hour. Let the glassware cool slowly in the oven after the oven is off before removing it. If the glassware has cotton or plastic tops, or contains fluids, it must be autoclaved in moist heat. Bottles with tops do not need to be wrapped completely in aluminum foil. Cotton stoppers should be covered in foil, and plastic tops should be screwed on only loosely. Once the autoclaving is completed, the screw caps are tightened.

Most researchers now use disposable, presterilized plastic dishes for culture chambers. If you can afford them, they are well worth the expense for the convenience. Be aware that they are virtually impossible to resterilize, but they can be reused for nonsterile purposes.

Fluids

Fluids can be sterilized by boiling, autoclaving, or filtering. The method used depends upon the solution.

Water can be boiled. Many solutions, however, cannot, since boiling will change their concentration. Then autoclaving in moist heat will often suffice. Bottles should be filled less than three-quarters full, and tops should be loose during the autoclaving. After autoclaving, when the fluids have cooled, the tops can be tightened.

Many solutions, such as physiological saline solutions, cannot tolerate either boiling or autoclaving. Under heat, various of the components, such as calcium chloride and sodium bicarbonate, combine to form a precipitate (in this case, calcium carbonate). In these situations the fluid can be sterilized by passing it through a filter with a porosity that excludes bacteria—a porosity of 0.22 μm or less. There are many sterilizing filter setups available. Some use filtration under vacuum for large quantities of fluids, and others attach to a syringe for smaller quantities. It is an extremely fast method of sterilizing a solution and very handy for last-minute needs. You must have a sterile bottle ready to store the fluid in, of course.

Rules for the road

1. *Never breathe directly over a culture dish.* Observe your cultures while they are covered, or from the side whenever possible.

2. If you need to observe your culture uncovered, keep the cover raised above the culture dish to protect it as much as possible from your breath and falling germs. If you must place the cover down, invert it before putting it on your work surface.

3. When not working on your culture, keep your culture dish covered.

4. *Never breathe directly into a solution bottle.* If you need to smell a solution, waft the vapors toward you with your hand.

5. When using a solution, remove what you need from the stock bottle by pouring or using a sterile, cotton-plugged pipette. Keep the top to the bottle in one hand, pointing downward; do not lay it down. Flame the opening to the bottle after use and immediately put the top back on.

Accompanying Materials

Tyler, M. S. and R. N. Kozlowski. 2000. *Vade Mecum: An Interactive Guide to Developmental Biology.* Sinauer Associates, Sunderland, MA. "Tools." This chapter of the CD illustrates, with movies and still pictures, how to make microknives, microneedles, hairloops, micropipettes, and embryo spoons. "Laboratory Safety." This module illustrates a set of rules that should be followed in the laboratory to ensure safety.

Selected Bibliography

Cameron, G. 1950. *Tissue Culture Technique.* Academic Press, New York. Because of the modern surge in expensive disposables, the art of the reusable is submerged in the older literature. This dated volume is a wonderful, compact addition to the low-cost lab and discusses space, equipment, and sterilization procedures.

New, D. A. T. 1966. *The Culture of Vertebrate Embryos.* Logos Press, London. It is hard to find a volume that discusses embryological tools. This book has a short chapter on several of the classics such as glass bridges, microneedles, and micropipettes.

Rugh, R. 1948. *Experimental Embryology: A Manual of Techniques and Procedures.* Burgess Publishing, Minneapolis, MN. This very old, out-of-print text is both a classic and a gold mine, and it should be bought whenever it is discovered on the dusty shelves of a used bookstore. An introductory chapter discusses some toolmaking techniques and is followed by a wealth of information on amphibian development and experiments.

Suppliers

Any good hardware store:

Dowels
Sheet metal
Aviation tin snips, dikes

Any good biological supply house, such as:
Fisher Scientific
585 Alpha Dr.
Pittsburgh, PA 15238
1-800-766-7000
www.fishersci.com

Syringe filters for sterilizing fluids; order 0.22-μm pore size
Vacuum filter units for sterilizing fluids; order 0.22-μm pore size
Culture dishes; a number of types are listed under petri dishes
Insect pins
Germicidal lamps (8 watt); they come in 12- through 36-inch lengths. General Electric
    makes these and distributes them through science and medical supply houses.

Fine Science Tools, Inc.
373-G Vintage Park Drive
Foster City, CA 94404
1-800-521-2109
www.finescience.com

This supply house carries very fine and expensive dissecting tools. They also carry excel-
    lent fine forceps (Dumont), graded as student forceps, that can be bought in bulk in-
    expensively. They also carry special razor blades that break easily and insect pins
    down to #000.

Nasco
901 Janesville Ave.
Fort Atkinson, WI 53538-0901
1-800-558-9595
www.nascofa.com

This supply house carries inexpensive fine forceps and dissecting tools, with additional
    savings when items are bought in bulk.

Any good craft or woodworking store, such as:
Tandy Leather Company
Outlets all over. Catalog from:
1400 Everman Pkway
Fort Worth, TX 76140
1-888-890-1611
www.tandyleather.com

Woodcraft
210 Wood County Industrial Park
P.O. Box 1686
Parkersburg, WV 26102-1686
1-800-225-1153
www.woodcraft.com

Arkansas stones and Japanese waterstones

chapter **3** *Using the Compound Microscope*

The Microscope

The compound microscope is an exquisitely refined, expensive piece of equipment. Learning how to use it correctly not only will allow you to achieve the greatest resolution your microscope can give, but it also will avoid costly damage. A microscope has a number of components, all of which you must be familiar with (see Figure 3.1). By the time you finish this lesson, not only will you be able to say "Koehler illumination" three times fast, but you will know what it is, how to use it, and why.

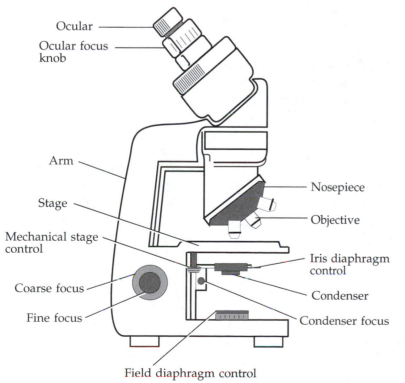

Ocular
Ocular focus knob
Arm
Stage
Mechanical stage control
Coarse focus
Fine focus
Field diaphragm control
Nosepiece
Objective
Iris diaphragm control
Condenser
Condenser focus

Figure 3.1
Schematic diagram of the compound microscope.

Safety first

First, learn how to carry the microscope. Always carry it by its **arm** (Figure 3.1), supporting the **base** with your hand. Once it is at your lab station, determine whether the light source is built in or is an external lamp. If it is built in, plug in the cord and briefly look at all parts of the scope. Find the on-off switch, which is usually built into the base and is often on a rheostat that allows you to vary the intensity of the light with the turn of a knob. If there is a meter to tell you how much light you are using, often this meter will have a red zone at the high end of its scale. Very seldom would you need to have the light on this brightly, and it is very hard on the bulb to do so, significantly reducing its life. Keep light settings low.

Stage

Examine the microscope's **stage**. Most stages have a clip for holding a microscope slide in place. This will be either a pair of stationary spring clips or a spring-loaded lever attached to mechanical controls that can be used to move the slide on the stage back and forth. Mechanical stage controls are usually two knobs set below the stage. Find these knobs and move the mechanical stage to learn which knob governs which direction of movement. Now place a microscope slide in place. If you have a spring-loaded slide holder, pull the lever back, slip the slide in place, and *very gently* release the lever. Many a slide is broken by an inadvertent clumsy release of this lever allowing the holder to snap against the slide.

If you have a mechanical stage, look at it again to see if it has two **vernier scales**—one along the top of the stage, the other along the side. As you move the stage, notice that the position of the slide is indicated by these scales. Vernier scales are particularly handy when working with **serial sections**—for example, when you want to keep track of particular sections out of the many that are on the slide. This is especially useful in photomicrography for recording precisely what objects have been photographed.

Magnifications

Now look at the **objectives** held in the nose piece. You probably have three or four. The magnification of each objective is written on its barrel. Find out which magnifications you have. Now look at the magnification of your **ocular(s)**. You have one ocular if your microscope is a monocular and two if it is binocular. The magnification of the ocular is either written on its rim or etched on its barrel. (Barrel markings can be seen by very carefully removing the ocular.) The magnification that you see looking through the microscope is the magnification of the objective multiplied by that of the ocular. Write down the magnifications that your microscope gives. If one of your objectives is 100× oil, then this is an oil-immersion lens and can be used only in combination with immersion oil. Never use this lens unless your instructor allows immersion oil to be used.

Oculars

Before using a binocular microscope, you must adjust the oculars so that the distance between them matches the distance between your eyes (your interpupilary distance). Usually you can do this simply by pulling the objectives away from each other or pushing them together. Now, rotate the nosepiece so that the lowest-power objective is in place. The microscope should always be stored with this objective in position. Find the **focus knobs** on your scope. There should be both a coarse- and a fine-focus. As long as the lowest-power objective is being used, you can start focusing by first adjusting the coarse-focus knob to move the objective as close to the slide as it will go. There will always be plenty of clearance between the slide and the lower lens of this objective. Make sure that the slide is positioned so that there actually is something in view; then, while looking through the ocular(s), use the coarse-focus knob to slowly move the objective farther from the slide. Objects on the slide will suddenly come into focus. You then use the fine-focus knob to adjust the final focus.

If your microscope is binocular, usually one of the oculars will be adjustable by means of a focus knob around its circumference. To set the oculars for your eyes, first fine-focus the microscope on a particular detail, using only the eye that is looking through the nonadjustable ocular. Then close that eye and look through the adjustable ocular with the other eye. Rotate the ocular's focus knob (not the microscope's focus knobs) until you again see the same detail in sharp focus. Now look with both eyes through the oculars. Everything should look crisp, and your eyes should feel relaxed. If you do not adjust the oculars to your eyes in this manner, in no time you will start feeling nauseous or have a pounding headache, as your eyes try to compensate for the out-of-focus ocular. It will take the fun out of the day. Sometimes a person feels slightly sick after using the microscope even when the oculars are properly adjusted. In this case, the person is unconsciously focusing images with their eyes rather than focusing the microscope. The trick is to *completely relax your eyes as you use the microscope; do all the focusing with the focusing knobs.*

Objectives

We now come back to the objectives. On most microscopes, these are parfocal with one another. "Parfocal" means that, once the specimen is in focus using the lowest power objective, all the other objectives will be roughly focused on the specimen when moved into position. A microscopist always starts focusing with the lowest-power objective in place. Only the sloppy novice bulls ahead using a higher-power objective first. Chide the sloppy novice, for they, in one click of a long-barreled objective, can crash that objective into the slide, cracking both the slide and the objective lens. This is a $400 to $1000 mistake. *Always start focusing with the lowest-power objective* and move progressively up through higher-power objectives, fine-focusing each one, until you reach the objective that you want. When moving from a lower-power to a higher-power objective, *always check from the side to make sure there is sufficient clearance for the objective*. A thick slide will never fit under a high-power objective.

The quality of the image you achieve with your microscope depends to a great extent on the **numerical aperture**, or **NA**, of your objectives and the degree to which the objectives have been corrected for aberrations inherent in lenses. The NA of a lens is a measure of its light-gathering ability—the wider the cone of rays that can be gathered, the higher the NA. And the higher the NA, the better the resolution of that objective. An oil-immersion lens can have a NA greater than 1 (up to 1.3), but all dry objectives have an NA of less than 1. Look at the barrel of each objective to find the number that corresponds to its NA and record this in your lab notebook. Which objective has the lowest NA? Which the highest? Which objective (other than the oil-immersion) gives the best resolution?

A microscope lens is subject to spherical and chromatic aberrations. In **spherical aberration**, the periphery of the lens focuses more strongly than its center. In this case, if you focus on the center of the image, things on the periphery are out of focus. In **chromatic aberration**, different colors are focused to different planes, causing fringes and halos of color in the image. Focus on a slide using each of your objectives, and determine for which of these aberrations, if any, your objectives have been corrected. Now check to see what is written on the side of each objective. (The type that it is will be abbreviated.) Chromatic aberrations can be corrected to various degrees. Achromat lenses are the least corrected, followed by semiapochromat (fluorite), and finally apochromat lenses, which are the most highly corrected for chromatic aberrations. Objectives that are highly corrected for spherical aberration have the word "plan," meaning flat-field, added to their name. These are top-of-the-line lenses and are highly recommended for photomicrography. A planachromat, therefore, is a very good lens, and a planapochromat is the Mercedes of all lenses. Record what type lenses you have.

Now that you are familiar with the basics, it is important that you learn how to achieve maximum resolution from your microscope. The **resolution** of a microscope is the precision with which you can use it to distinguish separate items. The closer two items can be and still have the microscope resolve them as two items instead of blurring them into one, the better the resolution of the microscope. The resolving power of most classroom microscopes is superb, regardless of how old they are. The physics of optics sets a limit to resolving power of light microscopes, and this limit was achieved by the end of the 1800s. By this time lens makers were able to correct to a great extent the aberrations in glass lenses. (Improvements in contrast techniques, however, continue to be made.) A good microscopist knows how to adjust the settings on the microscope to achieve the best resolution—that is, to set apertures and illumination so that the effects of aberrations in the lenses are minimized. This array of settings is known as **Koehler illumination**. (This very German word, "Koehler," is pronounced almost like the very American word "curler.")

Koehler Illumination

The condenser

Examine the parts of your microscope that lie below the stage. Most microscopes have a set of lenses here called the **condenser**. The condenser's job is to focus light from the light source onto the specimen. Look for a knob that when turned moves the condenser up and down. By moving the condenser, the focal plane of the light is moved. To make sure that it is focused right at the plane of the specimen, you must do the following. First, focus your microscope on an object. Then, if your light source is built in, see if there is also a **field diaphragm** (see Figure 3.1) on the base of the microscope that can be opened and stopped down to change the diameter of the cone of light coming from the light source. If you do not have a field diaphragm, you can make a temporary one with a heavy index card. Cut a circle of card so it can sit directly on the field lens, use a hole-punch to cut a central hole in the card, and center it on the field lens (where the light is coming through). If you have a field diaphragm, stop it all the way down. Now look through the microscope. You should see a field of light that is smaller than the full field of view; the edges of this field are probably fuzzy. This will be the case if the condenser is not properly positioned. Move the condenser up and down until the edges of the lit field come into sharp focus. ***Do not change the focus of the microscope during*** *this operation.* Remember, you are only moving the condenser up and down. When the edges of the light field are in sharp focus, the condenser is properly positioned to give you the maximum resolution for the particular objective you are using. Now check the position of the condenser. It should be fairly high, several millimeters from its highest position. The field diaphragm now can be opened back up until the cone of light just fills the field of view. If it is opened wider than this, you lose some resolution.

Iris diaphragm

The **iris diaphragm** must also be adjusted for best resolution. You should find it attached to the condenser (Figure 3.1); it is regulated by a lever. This diaphragm, when properly adjusted, excludes stray, refracted light that would interfere with sharp imaging. The only professional way to adjust this diaphragm is annoyingly time-consuming. You must remove an ocular and look down the barrel of the microscope. You are now looking at the back lens of the objective. Start with the iris diaphragm completely open, then close it down until you see that light is filling two-thirds of the back lens of the objective. Since the setting differs with each objective, the iris diaphragm must be readjusted every time a different objective is clicked into place. If you are doing photomicrography, it is well worth the time to set the iris diaphragm exactly, but it really isn't worth the time or the risk of dropping the ocular if you are simply studying slides. You can get a good approximate setting by closing down the iris diaphragm just to the point at which the illumination appears to dim. Get used to how objects look when the iris diaphragm is properly set. In general, if the objects look grainy, the iris diaphragm is closed down too far; if they look washed out, the iris diaphragm is open too wide.

You have now achieved Koehler bright-field illumination and the maximum resolution for your microscope. Congratulations.

Other Settings When Maximum Resolution Is Not Desired

There are times when you do not want maximum resolution from your microscope. Since resolution is gained at the expense of contrast, there may be times when you want to sacrifice resolution to increase contrast. In any type of microscopy, as well as in photography, if there is a need to increase contrast, it must be done at the expense of resolution. Resolution and contrast are inversely related. The higher the contrast, the lower the resolution. Often, specimens on slides are only faintly stained, or you are looking at unstained material. Under these circumstances, contrast must be increased in order to see anything. There are several ways to do this, which the lengthy discussion of Koehler illumination above should reveal to you. Try to discover them on your own before taking the easy way and reading about them below.

The first thing a microscopist tries when increased contrast is required is to stop down the iris diaphragm. This gives a very unpleasing grainy texture to the object being viewed, but dramatically increases contrast. Remember, if your specimen is already adequately stained, then stopping down the iris diaphragm seriously compromises the details that you can see.

A second method of increasing contrast is to jack the condenser down to a low level. Play around with your microscope settings, view the results, and record both what you did and the results in your laboratory notebook.

Oil Immersion

It is very seldom that you would have to use oil immersion. Your lab instructor may prefer that you not use it at all because of the difficulty of properly cleaning the objective lens and slide after use. An example of a specimen for which oil immersion would be useful is a slide of sperm.

It is exceedingly important that any slide you use for oil immersion is not a thick one. A whole-mount preparation of a chick embryo, for example, would be too thick. There is not enough clearance for the long 100× objective, and it can be damaged by contact with the slide. (Even a cheap oil-immersion lens can cost $150.)

You can tell if your microscope has an oil-immersion lens because "oil" or "öl" will be written on the barrel along with the magnification. Very fancy microscopes may have lower-power oil-immersion lenses, but most microscopes only have a 100× oil-immersion objective. A special oil, which has the same refractive index as glass, is used as a medium between the slide's coverslip and the objective. This oil increases the light-gathering power of the lens. With dry lenses, light rays have to pass through glass, into air, and back through glass, and at each interface some light is scattered. With immersion oil, the interfaces scatter less. Resolution is enhanced because more light from the specimen (more information) reaches the lens.

The first step in setting up an oil-immersion lens is to focus the specimen, starting on the lowest-power objective and working up to the 40× objective. Set the controls for Koehler illumination at 40×. Then swing the 40× objective out of position without changing the focal plane, put a drop of immersion oil on the slide precisely where the 40× objective had been, and swing the oil-immersion objective into place. The oil should be caught by the lens of the objective and fill the tiny space between coverslip and lens. If there is still an airspace, add a second small drop from the side. Now focus the specimen using the *fine-focus only*. Remember, your objective lens is *very* close to the coverslip, and a slight movement of the coarse-focus could bring the lens crashing into the coverslip. You can now reset the settings for Koehler illumination at 100×.

You can move the slide around with the mechanical stage. If you don't have a mechanical stage, moving the slide will be difficult, since a slight movement with the hand will register as an enormous distance under the lens. If you do move the slide, keep checking to make sure that there is still enough oil between the lens and the coverslip.

Cleaning up after using oil immersion is critical. If oil is left to dry on the lens, it will take a professional technician to clean it adequately. Use lens paper designed specifically for microscopes. (Never use lens paper for eyeglasses, for example. It is coated with a compound that comes off on the lens and is a trial to remove.) Wipe the lens once in one direction, move to a clean area of lens paper, and wipe the lens again. Repeat this until no more oil shows on the lens paper. Clean the oil from the slide as well. In this case, Kimwipes™ can be used, and even a small amount of soapy water, if necessary.

A note of caution: In the past, a type of immersion oil was routinely used that contained PCBs (polychlorinated biphenyls), which are both carcinogenic (cancer-causing) and estrogenic (acts as an estrogen). New federal regulations require that only PCB-free immersion oil be used, but occasionally an old bottle shows up. Check the immersion oil you are using. Make sure that its label says "safe" or "contains no PCBs."

Make Your $1000 Microscope into a $10,000 Instrument for Pennies

Color filters

Color filters can be used to increase resolution and contrast. They are extremely useful both for photomicrography and for studying a specimen. You can make a $100 set of filters for approximately $5.

Obtain a set of differently colored pieces of cellophane, each approximately 2 inches square. The most useful colors will be red, blue, green, and yellow. The cellophane squares can be mounted in cardboard mounts. These can easily be made from posterboard, or you can use premade mounts that are used for projection slides. Before mounting the cellophane, cut the opening in the cardboard mount so that it is as large as possible. Cut the square of cellophane so that a narrow border of cardboard is left around the rim that isn't covered by the cellophane. Glue or tape the edges closed. Projection slide mounts often are preglued, and an iron is used to close these. You can use an iron specifically made for mounting slides or a regular household iron on a very low setting. Iron only the edges; do not let the iron slide over or touch the cellophane.

The filters are used by placing them on the field diaphragm. To increase resolution, use filters with shorter visible wavelengths. Within the visible spectrum, red has the longest wavelength (700 nm); orange, yellow, green, and blue are progressively shorter, in that order; and violet has the shortest (400 nm). Put a slide of a specimen on your microscope. Use each filter you've made, and note in your laboratory notebook any differences in resolution. It is helpful to focus on something such as cilia, which are near the limits of resolution of your microscope. You may find that you see best when using a green filter. Even though blue and violet have shorter wavelengths and therefore give better resolution to the microscope, your eye is most sensitive to green. It points out that, ultimately, resolution is a function of both the microscope and your eye.

You can also use color filters to increase contrast in a stained section. This is done by using the filter that is the complementary color to the color of the section. A red-stained tissue will look much darker through a green filter, for example. Red and green are complementary, as are blue and orange. Find slides that will allow you to test each one of your filters, and record your observations in your laboratory notebook.

Polarizing filters

You can make your microscope into a polarizing microscope very easily using polarized film. The polarization system that you can make for less than $2 costs over $500 from a microscope supply company. Polarized film transmits light that is vibrating only in one direction—all other rays are excluded. You will use this property in several applications.

Cut the polarized film to make two 2-inch squares. If the polarized film comes with a plastic protective sheet, this can be removed. Etch an arrow into the plastic film to show the direction of light transmittance through each polarizer. You can determine this by looking through the filter at a glare surface, such as a puddle with sunlight reflecting from it. Turn the filter until the glare is eliminated. The direction of polarization in your filter is now at right angles to the glare surface.

Check your polarized sunglasses in the same way. In which orientation would you expect the polarized lenses to be mounted?

Using polarized light will increase the resolution of your microscope. Put one of your polarizers on the field diaphragm, and focus on a slide. Do you notice an increase in resolution? How does it compare to using a blue filter? Combine both a blue filter and a polarizer. What do you notice? Record your observations in your laboratory notebook.

Polarized light can also be used to detect birefringence. **Birefringence** is the ability of a substance to rotate polarized light and can be used as a diagnostic characteristic of a structure. The highly keratinized layers of skin, for example, are birefringent. (Another word for birefringent is "anisotropic.") To check for the presence of birefringence, you need two polarizers. One is placed on the field diaphragm, and the other is placed over the eyepiece. The polarizers must be rotated so that they are crossed; that is, their directions of transmittance are at right angles to one another. The easiest way to determine if your polarizers are crossed is to rotate them until the field of view is dark. The light that passed through the first polarizer is blocked out by the second polarizer (called the analyzer). Anything between the two polarizers that is birefringent will rotate light coming through the first polarizer, and this rotated light can now pass through the second polarizer. The effect you see is a shining object against a dark field. Highly parallel arrays of closely spaced, linear structures are birefringent.

Focus on a slide of adult mammalian skin with hair follicles. Now position your two polarizers so that they are crossed. One should be on the field diaphragm, and the other on the eyepiece. If you do not see any birefringence, rotate the two polarizers so that they maintain their relative orientation to one another until you see the keratinized layers of skin shining. On an actual polarizing microscope, the stage is able to rotate so that you don't have to rotate the polarizers. Rotating the object in relation to the crossed polarizers maximizes the amount of light that is transmitted. Try slides of other material. Diatoms are exquisite under polarized light. Other good examples of birefringent substances are cellulose, starch, and mitotic spindles. Record your findings in your laboratory notebook.

Dark-field optics

Dark-field optics can dramatically increase contrast. By placing an opaque disk in the middle of the field lens, a hollow cone of light can be made that illuminates the object but does not enter the objective. Any light that is refracted by a structure will enter the objective and make its image appear self-luminous against a dark field. Dark-field is often used on living material such as ciliated organisms. Since cilia and flagella are near the limits of resolution of your light microscope, they are nearly invisible in bright-field, but the severe contrast that dark-field introduces will make them stand out sharply. Even objects smaller than the resolution of the light microscope, such as microtubules and bacterial flagella, can be seen in dark-field microscopy.

To achieve dark-field, you need only to create a hollow cone of light that has a diameter appropriate for the objective you are using. A simple way to do this is to center coins on the lens of the field diaphragm—a quarter for the lowest power objective, a nickel for a higher power, and a dime for the highest. If the diameter is correct, when looking through the ocular you will see just a bare fringe of light around the edge. Try looking at several samples of living material—a scraping from the inside of your cheek, a protozoan culture, a fly wing. Record your observations in your laboratory notebook.

Measurement under the microscope

The device normally used to measure structures under the microscope is a **micrometer**. This can be made as a slide or as an insert for the ocular. Micrometers are relatively expensive (about $30), consisting of finely etched glass showing several millimeters, each divided usually into tenths or hundreths. You can make a crude slide micrometer with a glass slide, an etching tool, or a photocopying machine.

Get a glass slide, a diamond- or carbide-tip marking pen, and a millimeter ruler. Work under a dissecting scope. Place the slide on top of the ruler under the dissecting scope and etch several centimeters with their millimeter markings into the glass. In the same manner, make a grid micrometer by etching the outline of a square centimeter and dividing it up into millimeter squares.

To use your micrometer, place it upside down over a slide that is already in focus. This puts the markings as close to the coverslip as possible. Make sure that there is clearance for this rather bulky prep, and only use the lowest two powers of your microscope.

A thinner micrometer can be made by photocopying a millimeter ruler and grid onto a transparency and cutting it out. This makes a very thin preparation, and is therefore preferable to the slide micrometer, but it cannot be sterilized.

With your micrometer and grid, you will be able to measure the diameters of structures in millimeters and to estimate subfractions of millimeters. Your micrometer and grid will be especially useful under the dissecting scope. Using both your dissecting scope and compound microscope, measure the thickness of the epidermis on the skin slide. Choose something else to measure under the dissecting scope. Record your measurements in your laboratory manual.

Care and Maintenance of Your Microscope: "Twelve Good Rules"

"'The time has come,' the Walrus said, 'to talk of many things.'" And those things for us are a most annoying, but most necessary, list of dos and don'ts that will make you a safe microscopist rather than a liability.

1. **Always** carry the microscope with one hand grasping the arm and one hand supporting the base, keeping the microscope level.
2. **Never** increase the light so that the meter shows in the red zone. If brief incursions into the red zone are necessary (for photomicrography, for example), keep them short, reducing the light as soon as possible.
3. **Never** leave the light on when the microscope is unattended.
4. **Never** force a knob or control. Get help from the instructor if something won't move.
5. **Always** move from the lowest-power objective to successively higher-power objectives.

6. **Never** unscrew an objective unless instructed to do so. If you remove an ocular for adjusting Koehler illumination, do not let go of it, and replace it immediately after the adjustment is made. Dirt that gets down the ocular tube impedes microscopy and can be removed only by a technician.

7. **Always** clean an oil-immersion lens immediately after use and clean any oil that might have spilled on the stage. Never use alcohols when cleaning; these will dissolve the glue that holds the lenses in place. A small amount of xylene or toluene may be used, but always check with your instructor first.

8. **Never** leave a slide on the stage when storing the microscope.

9. **Always** put the lowest-power objective back into place for storing the microscope.

10. **Always** cover the microscope when storing it. If there is no microscope cover available, get a plastic bag large enough to come down over the scope to a level lower than the stage so that the condenser is adequately covered.

11. **Always** review these rules until they are reflex.

12. **Always** take pride in being a fine microscopist, and keep the naive show-offs away from your precision instrument.

Accompanying Materials

Tyler, M. S. and R. N. Kozlowski. 2000. *Vade Mecum: An Interactive Guide to Developmental Biology*. Sinauer Associates, Sunderland, MA. "Microscope." This chapter of the CD gives step-by-step instruction on how to achieve Koehler illumination and how to use color and polarizer filters, and it shows the results that are seen through the microscope. A virtual microscope is also shown in a QuickTime Virtual Reality (QTVR) video in which the parts of the microscope are identified and defined.

Selected Bibliography

Bradbury, S. 1980. Getting the best out of your micrscope. *Royal Microscopical Society Proceedings* 15: 270–279. A concise and valuable set of tips and explanations.

Davidson, M. W. 1991. A photography primer. *The Science Teacher* 58(7): 12–18. This nontechnical, well-illustrated paper gives practical information about adapting the microscope for polarized and reflected light as well as photography.

Delly, J. G. 1988. *Photography through the Microscope*. Eastman Kodak Company, Rochester, NY. This is a beautifully illustrated pamphlet that concentrates primarily on photomicrography but has an excellent discussion of the principles of light microscopy, Koehler illumination, filters, dark-field, and polarizing methods.

Gage, S. H. 1920. *The Microscope*. Comstock Publishing Co., Ithaca, NY. Originally published in 1908, this book is a testament to how sophisticated microscopes were even in the early 1900s. Packed with information about microscopy and histological preparations, it is still extremely useful.

Lacey, A. J. (ed.). 1989. *Light Microscopy in Biology: A Practical Approach*. IRL Press/Oxford University Press, New York. This is a very sophisticated, but readable, volume that includes an such excellent chapter on the principles of light microscopy. Other chapters are on specialized techniques as immuno-chemistry, fluorescence, and video microscopy.

Needham, G. H. 1968. *The Microscope: A Practical Guide*. Charles C. Thomas, Springfield, IL. This older volume has lost none of its usefulness. The illustrations are clear, the language straightforward, and the basic information will always be relevant.

Sluder, G. and D. E. Wolf (eds.). 1998. *Methods in Cell Biology. Vol. 56, Video Microscopy*. Academic Press, San Diego. A sophisticated volume that helps make sense of terms and techniques in microscopy.

Suppliers

Edmund Scientific Company
101 East Gloucester Pike
Barrington, NJ 08007-1380
1-800-728-6999
www.edsci.com

Economical scientific equipment and gadgetry
Color filters (listed under photography); filters can be bought as individual sheets or a book of 6
Polarizing sheets (listed under optics), including an inexpensive, experimental grade

Any good photography store:
Cardboard mounts for projections slides (e.g., Kodak Ready-Mounts)

Any good scientific supply company, such as:
Ward's Natural Science Establishment, Inc.
P.O. Box 92912
5100 West Henrietta Road
Rochester, NY 14692-9012
1-800-962-2660
www.wardsci.com

Diamond- or carbide-tip marking pen. (A carbide-tip drill bit from any hardware store is an inexpensive alternative to the more expensive marking pens. For use, wrap all but the tip of the drill bit in tape to protect fingers.)
Immersion oil

4 Cellular Slime Molds
Mycetozoa:
Dictyostelium discoideum

Life Cycle

Scuffing through the moist leaves on a woodland path, you might have disturbed the tiny myxamoebae of the quietly present and seldom seen cellular slime molds of the group Mycetozoa. **Myxamoebae** are the vegetative stage of these primitive eukaryotes. Classified among the protozoa, cellular slime molds—maligned by their name—are neither slimy nor molds. Their vegetative stage is just the beginning of a simple and fascinating life cycle. This simplicity is one of the reasons for studying these organisms. With only a limited number of cell types, development and behaviors can be examined at the cellular, biochemical, and even genetic levels. The species of Mycetozoa most widely used for such studies is *Dictyostelium discoideum*.

The vegetative stage

The life cycle of *Dictyostelium* (Figure 4.1) starts with the hatching of myxamoebae from the cellulose encasements of the spores. Under favorable conditions of moisture and warmth, each spore releases a single, haploid ($N = 6$) myxamoeba (or simply, amoeba). These amoebae crawl on the substrate, feeding on bacteria and dividing by mitosis. In nature, this **vegetative stage** is probably the stage in which the organism spends most of its time. The amoebae move toward the bacteria, attracted by the **folic acid** the bacteria secrete. When the food supply is depleted, though, a second stage begins, that of **aggregation**.

Aggregation

Upon starvation, the amoebae go into a 4- to 8-hour refractory period when nothing appears to be happening. Using the fine resolution of a biochemist's magnifier, however, we see that there is an extraordinary list of activities. Starvation triggers a burst of gene activity in the amoebae. Possibly as many as 100 genes are called into action as a complex of new biochemical machinery is assembled: a set of **glycoproteins** that will enable cell–cell adhesions appear in the cell membrane; **adenylyl cyclase**, an enzyme that catalyses the formation of **cyclic AMP** (cAMP; adenosine 3´, 5´-cyclic monophosphate) from ATP is formed; **cAMP receptors** are put into the cell membrane; and a **membrane-bound phosphodiesterase** that breaks down cAMP appears, as well as a **phosphodiesterase inhibitor**. How will

Figure 4.1
Life cycle of the cellular slime mold, *Dictyostelium discoideum*. *D. discoideum* spends much of its life cycle as feeding myxamoebae on the forest floor. When food runs out, the myxamoebae aggregate to form a migratory pseudoplasmodium (or slug), which searches for a new environment that will be suitable as a feeding ground. Once this is found, the slug settles on its posterior end, taking a "Mexican hat" form, which then culminates into a mature fruiting body. The spores from the fruiting body are released and hatch into myxamoebae, starting the cycle over again.

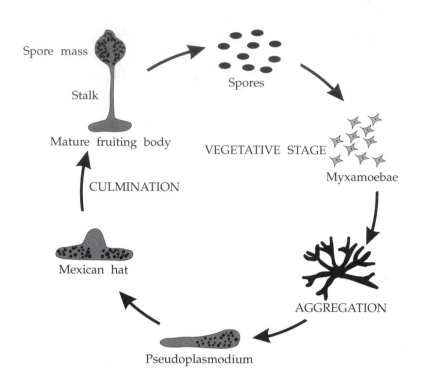

all this biochemical machinery be used? Surely something dramatic is about to happen.

Suddenly, directed cell movement begins. Amoebae within a certain area start migrating toward a single central amoeba, as if that amoeba had shouted, "OK, on the count of three. . . ." That call to aggregation turns out to be a **chemical signal** that is secreted first by the central amoeba, and then by all the amoebae in the area. The signal is **cAMP**. This is the same molecule (often called a second messenger) made by your own cells in response to hormones such as adrenalin, your morning coffee, or your asthma medication. In all of these cases, cAMP is made by adenylyl cyclase and kept within the cell. In the case of *Dictyostelium* aggregation, however, intracellular cAMP levels first rise by induction of adenylyl cyclase, and then cAMP pours out of the cell. This use of cAMP in the extracellular environment has never been demonstrated in any other organism.

The extracellular cAMP acts as a chemotactic agent, and the amoebae move in response to its presence, traveling up its concentration gradient. This is one of the clearest, most experimentally accessible examples of **chemotaxis** that we as developmental biologists have. With all amoebae pulsing out cAMP, a concentration gradient is established that is greatest at the center of a group of amoebae. The cAMP binds to the cell surface **cAMP receptors** of the amoebae, and this induces the amoebae to move in the direction of a higher concentration. When the cAMP-receptor sites are all filled so that there is no distinction between the front and back of the cell, an amoeba can no longer sense the gradient, and movement stops. The high levels of bound cAMP activate the **phosphodiesterase**, which breaks down the cAMP into 5′-AMP, thereby freeing the receptor sites. A **phosphodiesterase inhibitor** that turns off phosphodiesterase ensures that there is never too much phosphodiesterase made.

The cycle of cAMP secretion and amoeboid movement is repeated about once every 6 minutes. An amoeba pulses out cAMP and moves up the cAMP gradient for about 60 seconds, then remains stationary until the next pulse.

As the cells move toward the central amoebae, they bump into one another, and because of newly made **glycoprotein adhesion molecules** in their cell membranes, they adhere and form clumps. This results in streams of amoebae, traveling together toward the center. You will see this in your dishes as beautiful branched patterns radiating out from a central mound.

There are dozens of questions you can ask about this stage. For example, what would happen if you flooded the substrate with cAMP? If you increased the amount of phosphodiesterase?

Pseudoplasmodium

Aggregation takes from 8 to 12 hours from the time of starvation. The assembly of amoebae first form a mound, which elongates and finally tips over, becoming a finger-shaped **pseudoplasmodium**, **grex**, or slug (Figure 4.2). The slug, usually 2–4 mm long, can consist of only a few hundred or as many as 100,000 cells. It is motile, has anterior and posterior ends, moves only forward, and is attracted toward light, higher temperatures, and higher humidity—conditions that would attract it to the surface in its natural environment. In your own dish, you may see it wandering, seemingly aimlessly, crossing and recrossing its own tracks. You will be able to see a complete record of its wanderings, since the slug lays down a trail—a cellulose sheath—that is easily visible. This sheath, secreted by anterior cells, provides a sleeve through which the slug moves and which collapses in its wake.

Though the slug may look relatively undifferentiated at first glance, it contains several cell types. The anterior cells, constituting about 20% of the slug, are **prestalk cells**. They will form the stalk in the mature fruiting body. The posterior cells of the slug are primarily **prespore cells**, which become the spores of the fruiting body. Scattered throughout the posterior region is a third cell type, discovered quite recently, called **anterior-like cells**, which become the basal disc of the fruiting body and the upper and lower caps of the spore mass. Many have asked what the factors are that determine these cell types. As in the development of higher organisms, these sophisticated questions of induction and differentiation can be asked of this simple organism. Not all answers are in, but what is known is tantalizing. Cyclic AMP, as you might have suspected, continues to play a major role. High levels of cAMP appear to induce prespore cells, and a morphogen known as **DIF** (differentiation-inducing factor), a low molecular weight lipid found in the anterior region of the slug, determines prestalk cells. Little is known about the

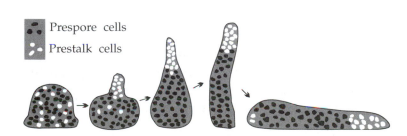

Prespore cells
Prestalk cells

Figure 4.2
Formation of the pseudoplasmodium. After aggregation, prestalk and prespore cells arrange themselves into a slug by first forming a mound, which elongates and finally tips over, becoming a migrating pseudoplasmodium. The prestalk cells position themselves anteriorly, the prespore cells posteriorly.

anterior-like cells. Surprisingly, cells may already be determined prior to aggregation, and their regional separation in the slug may be the result of sorting out of the cell types.

One of the major questions of developmental biology is whether the genetic programs of differentiated cells can be changed or even reversed. This is something you easily can ask of *Dictyostelium* in the laboratory. Suppose you were to divide the slug into separate pieces. Would the cells of each piece form only what they were originally destined to form, or would they **regulate** to form normal fruiting bodies? What would happen if you disaggregated the cells of the slug—would they reaggregate? If you provided them with bacteria, would they revert to a feeding stage? How stable or labile are these cell types?

Culmination

Transformation of the slug into a **fruiting body** starts when the slug settles onto its posterior end, looking very much like a **Mexican hat** (and so it is called), and the cells start to reorganize. You might have guessed already that this would have to happen. The cells in the slug are positioned inappropriately for fruiting body formation. The prespore and prestalk cells will have to switch places. This starts as the anterior tip of the Mexican hat begins secreting a cellulose tube. Cells just behind the tip start migrating up the outside of the tube and into its open end. More posterior cells then start their journey up the sides of the tube, and the tube continues to be laid down and lengthens. In this way, the order of the cells is reversed. The anterior prestalk cells of the slug now form the stalk of the fruiting body, and the more posterior prespore cells of the slug become the spore mass of the fruiting body. This reorganization takes approximately 8–10 hours and results in a beautifully delicate **mature fruiting body**, usually 1–2 mm high. The stalk cells, now dead, have raised the spore mass high above the substrate. Poised to start the cycle over again, it awaits the proper stimulus to release its spores to the substrate.

The fruiting body is a stage that you will want to watch carefully, since it will give you important information about previous experiments. If, for example, you had cut a slug in half and each half formed a fruiting body, the relative sizes of the stalk and spore mass will tell you whether cells were reprogrammed to become different cell types. Also, you can manipulate different stages from Mexican hat to mature fruiting body to ask questions about regulation in these cells. Remember, however, that stalk cells die, and dead cells won't reprogram.

Sexual reproduction: A rare event

Though normally reproducing asexually, *Dictyostelium* can reproduce sexually under conditions such as excessive water and darkness. For sexual reproduction, different mating types must be present. The cycle begins in the aggregation stage when two amoebae in an aggregate fuse, creating a **giant cell**. This giant cell engulfs the surrounding cells of the aggregate, then encysts in a thick cellulose wall forming a **macrocyst**, which is resistant to harsh conditions. Within the macrocyst, the giant cell divides meiotically, becoming haploid, and then divides mitotically, producing a number of amoebae that eventually are released as feeding amoebae. Sexual reproduction is rare in these organisms, and it is not always successful. For example, of those macrocysts formed by *Dictyostelium discoideum* that have been observed in the laboratory, none have been seen to germinate.

Preparing for Your Laboratory Studies

The advantages to using *Dictyostelium discoideum* for our studies are many. The organism is extremely resistant to damage. You can hit it with a hammer and find that you've merely asked an interesting question: "What happens when the organism is disaggregated?" Also, the organism's simplicity invites us to examine complex principles in the context of a minimum of cell types. Developmental biologists have recognized these advantages for years and have used *Dictyostelium* to ask questions about mechanisms of cell movement and chemotaxis, about the biochemistry of differentiation, and about the genes involved in differentiation and pattern formation. Though genetic manipulations are beyond the scope of this laboratory, many other realms are open to you. Use your imagination. Design simple experiments that will yield sophisticated answers. Keep in mind that a chemical usually has more than one effect. And enjoy this experimental playground.

Culture procedures

The following is a conventional method for culturing slime molds. Petri dishes may already have been prepared for you with **nutrient agar**, a culture medium described below. A divided petri dish with four wells of nutrient agar makes an excellent set of experiment chambers. If your petri dish is not divided, you can make barriers with sterilized (use 70% alcohol or boiling water) coverslips pushed into the agar. The agar should have a moist but not excessively wet surface. Use sterile distilled water to wet the surface if needed. This culture medium will provide nutrient for *Escherichia coli*, the bacterium upon which the slime mold amoebae will feed. It is a substrate that is perfectly adequate for all stages of the slime mold's life cycle. Wherever the amoebae clean out an area of bacteria, they will aggregate to form slugs, and eventually there will be a rich population of slugs and fruiting bodies in the dish.

Nutrient agar culture for **E. coli**

Bacto-peptone	10.0 gm	*Buffer to pH 6.6 (usually doesn't need*
Bacto-dextrose	10.0 gm	*adjustment). The medium can be*
$Na_2HPO_4 \cdot 12H_2O$	0.96 gm	*autoclaved, or boiled for 3–5 minutes,*
KH_2PO_4	1.45 gm	*then poured into petri dishes. Recipe*
Bacto-agar	20.0 gm	*is enough for pouring about 35 plates.*
Distilled water	1000 ml	

Sterilize a bacteriological loop in a flame by heating it until it is red-hot. Allow it to cool, then immediately use it to transfer *E. coli* to the first well or small region of your agar plate. This is done by scraping some bacteria from the surface of the stock tube onto the loop and then lightly moving the loop back and forth over the surface of the agar in your plate. This will create streaks of bacteria. (A suspension of bacteria in sterile distilled water also can be used and smoothed over the surface of the agar with a bent glass rod. This will give an even lawn of bacteria rather than streaks.) Resterilize the bacteriological loop and use it in the same manner as above to transfer mature slime mold fruiting bodies and spores to the same well or small region of the petri dish where the bacteria have been placed. Label the dish with your name and date, and place the dish in a protected area of the lab. The optimum temperature range for growth of the slime mold is 22–24°C (room temperature) and optimum relative humidity is 70–75%.

The spores will germinate, releasing myxamoebae that will increase by mitosis and feed on the *E. coli*. This will continue as long as the supply of *E. coli* remains. Check on your dish as often as time allows during the week, keeping a precise record of the development you observe. Diagram what you see, note variations among different groups: What are the sizes of the aggregating populations that you see? What sizes of slugs do you observe? What different morphologies do you see among fruiting bodies? By the next week, all stages of development should be available to you in your culture dish, since this method does not produce synchronized development. You will do the rest of this lab at that time.

If the laboratory instructors have already prepared cultures for you, you will be starting your laboratory exercise at this point. When you receive your dish, answer the questions stated in the paragraph above.

Make as many observations as you can about the organism before proceeding to the experimental exercises. Examine different stages under the dissecting scope. Put various stages on a slide, and examine them under the compound microscope. Put a spore mass on a slide, and use a needle to burst it open. Look at the spores under high power; there will be about 70,000 in one spore mass. Each spore is ellipsoid and about 5 μm long. Make diagrams of what you see.

A note to the instructor: Preparing cultures

The most economical way of preparing cultures for class is to make subcultures from stock tubes bought from a biological supply house. In this way, just a few tubes of *D. discoideum* and *E. coli* can supply enough material for even a hundred students. Timing is the critical factor. If the class is small, making subcultures 5 days in advance of the class and keeping them at room temperature should work well. If the class is large, it is best to make subcultures 13 days prior to class. Use these to seed further cultures for the individual students 5 days in advance of class.

Experiments Using Dictyostelium

You will be designing your own experiments. What follows are suggestions of techniques you can use. When you start an experiment, state clearly in the "Experimental Setup" section of your laboratory notebook the question or questions you are asking. This is absolutely essential. You need to know what you are asking before you can know what the answers are. In addition to your observations of normal development, it is suggested that you do three experiments.

Labeling with vital dyes

You can mark individual organisms or regions of organisms with vital dyes. A **vital dye** is one that is not toxic to cells when used in low concentrations. It is toxic at high concentrations, however; the rule of thumb is that a vital dye should not be used at concentrations higher than 0.1%. This is considered a stock solution. The solution must be sterilized before use. This can be done by filtering it through a 0.22-μm porosity filter into a sterile test tube. When choosing a dye, you should always be aware of what elements in the cell the dye stains. Often valuable biochemical information can be gained by noticing the pattern of staining. Useful vital dyes include Nile blue sulfate, neutral red, methylene blue, and toluidine blue (Table 4.1).

You can use a vital dye to stain any of the stages in the life cycle of the slime mold. You could stain one group of amoebae and follow them through the rest of the life cycle. You could stain a slug and look for differential staining between prestalk and prespore cells. Move the structure you are staining to an unused quadrant of your culture dish (unless you are staining aggregating amoebae, which should be left in place so as not to disturb their configuration—unless a disturbance is part of the experimental design).

Place a drop of dye directly on the cells you are staining. Excess dye will sink into the agar and doesn't need to be removed. Record which stain you used. Diagram the staining pattern. If you used a metachromatic dye, note regions of primary and secondary colors. What conclusions about biochemical differences can you draw from the differences you see? Observe the organism throughout the rest of its life cycle, noting any changes in color patterns.

Table 4.1 Some useful vital dyes

Vital dye	Some properties
Nile blue sulfate	More or less specific for phospholipids and lipoproteins of cell membranes
Neutral red	Beautiful dye with variable staining properties that are difficult to interpret chemically. Stains autophagic vacuoles in prestalk cells of *D. discoideum*
Methylene blue	A basic dye. It is a "redox" dye–it is blue when oxidized and changes to colorless when reduced.
Toluidine blue	A basic dye. Extremely useful for its property of metachromasia, which means it has more that one color form. Its primary color, blue, shows when the dye binds to nucleic acids (DNA and RNA). Hence, the chromosomes in the nucleus will stain blue, as will ribosomes and other RNA-containing structures in the cytoplasm. The secondary color, pink, occurs when the stain binds to glycoproteins and mucins (mucins are not found in *D. discoideum*)

Transecting, grafting, and disaggregating pseudoplasmodia

Pseudoplasmodia (slugs) can be transected (cut) into halves, thirds, and so on. They also can be disaggregated into individual cells. A number of interesting questions can be asked in this way. For example, are the fates of the cells of a slug irrevocably set? If a slug is cut in half, what will happen? Can a half-slug survive? Can it differentiate? If so, what will it form? Will the anterior half form only stalk, or will some of the cells alter their fate to become spore cells, thus forming an entire fruiting body? If so, is the fruiting body that results normal in its proportions? Do the results vary depending upon how much time the slug wanders before it settles down to differentiate? What happens if you cut off just the tip of a slug? What do your results say about the flexibility of cells whose fates have already been determined? *The more questions you ask of a single experiment, the more acute your observations will be, and the more data you will collect from each experiment.*

To divide a slug into two or more pieces, first move the slug to an unused quadrant of your culture dish. This can be done by lifting the slug on the flat side of a microknife, or even by lifting out a section of agar along with the slug so as not to risk hurting the slug. You will have an easier time cutting the slug if it is thoroughly cooled first. Place your dish in the refrigerator for 10 minutes, or place it on crushed ice held in a larger dish. Use a microknife to cut the slug. Cuts can be made through the slug into the agar below. The knife is then drawn through the agar without touching the cells. Always make careful diagrams of the position of your cuts. Devise a scheme that will allow you to distinguish one piece from another. For example, each piece could be stained with a different vital dye.

Cutting the slug often results in bursting the cellulose sheath that encloses the cells, and instead of two discrete pieces you have a puddle of disaggregated slug. This is fine—don't despair. Your experiment has simply changed to being a disaggregation experiment. What will happen when you disaggregate a slug? Will the cells reaggregate? If so, will they reaggregate immediately, or will there be a lag period prior to reaggregation? What does this mean at the biochemical level? Will the disaggregated cells behave differently if you give them food (*E. coli*)? Can you force the life cycle to go backward? Just how flexible is a cell that has already become differentiated as a prespore or prestalk cell?

Grafting one piece of a slug onto another is rather more difficult than a simple cut, but can be used to ask more complex questions. What would happen if two anterior ends were grafted together? Would there be a battle for anterior dominance? In which direction would the resulting slug move? What would it develop into? The difficult part of grafting is keeping the two pieces together until they have annealed (joined together). It will help if you cool the slugs before cutting. It will also help if the slugs to be cut are close together initially, which diminishes the distance each piece must be moved. Choose the pieces you want to anneal, transfer one to the other by lifting it on the flat side of a microknife or other suitable instrument, and gently push the two pieces together. It will help to keep the dish cool during the first few hours of annealing. You could also dig a shallow depression in the agar where the pieces could sit. The shape of the depression should be such that it helps to force the pieces together. Use diagrams to indicate exactly the regions that were used in the graft.

Disruption of cyclic AMP levels

Altering cAMP levels will undoubtedly have profound effects on the organism. It is important to know that exogenous (extracellular) cAMP will not cross cell membranes to enter a cell and raise intracellular cAMP levels.

There are various chemicals that you could use to disrupt extracellular levels of cAMP. You could use cAMP itself, soaked into an agar cube or flooding an entire area of agar. The concentration of cAMP you use is your choice, but you should at least know some effective ballpark figures. Aggregating amoebae can respond to cAMP levels as low as 0.01 µg/ml. Levels used in chemotaxis experiments are often much higher than this, ranging from 0.05 to 1.2 mg/ml. At 1.2 mg/ml, amoebae fail to continue development.

Since cAMP may be something your laboratory cannot supply, there are other sources readily available. The tip of the migrating slug, for example, secretes cAMP and can be removed and placed among amoebae. Also, human urine has relatively high levels of cAMP (approximately 0.53–0.75 µg/ml), well within the range that amoebae respond to. Urine, of course, is a complex substance, and this must be kept in mind when you are drawing your conclusions. For example, urine also contains cGMP (approximately 0.09–0.23 µg/ml), which acts to block the cells' phosphodiesterase. (This should actually augment the rise in extracellular cAMP levels.) If you use urine, measure its pH (on average, this will be pH 6.3, but normal ranges are from pH 4.5 to pH 8.0), and be aware that it contains some ammonia (0.07–0.13% on average). Basic pH and ammonia are known to favor spore formation and decrease stalk formation.

Another method of altering cAMP levels is indirectly, by stimulating or inhibiting the enzymes involved in making cAMP. *Dictyostelium*'s phosphodiesterase, the enzyme that breaks down cAMP, is secreted at low levels during the amoeboid stage, and it is found at much higher levels, inserted in the membrane just prior to aggregation. During the amoeboid stage, the extracellular phosphodiesterase probably is useful in breaking down the cAMP secreted by bacteria. This would prevent the amoebae from responding prematurely to cAMP. During aggregation, the membrane phosphodiesterase cleans off cAMP binding sites, getting them ready for each new pulse of cAMP. What do you think would happen if you stimulated or inhibited phosphodiesterase during aggregation?

In higher organisms, caffeine or theophylline could be used to inhibit phosphodiesterase, thereby raising cAMP levels, but the phosphodiesterase of *Dictyostelium* is relatively unresponsive to these substances. Caffeine at a concentration of 10 mM (0.194 mg/ml) does not inhibit it, and theophylline will decrease its activity by 15% when present at a concentration of 10 mM (1.8 mg/ml). In fact, caffeine's effect on *Dictyostelium* is just the opposite of what you might expect. It actually indirectly increases levels of phosphodiesterase activity by inhibiting adenylyl cyclase. Intracellular cAMP levels drop, and this, in turn, may raise phosphodiesterase activity. 5 mM caffeine (0.97 mg/ml) can double the activity levels of phosphodiesterase in *Dictyostelium* amoebae.

If you choose to use caffeine or theophylline, you can use the purified form, or a cheaper source—that cup of coffee or tea you're not allowed to drink in the laboratory. Drinking strengths of coffee contain 0.5 to 1 mg/ml of caffeine, and a cup of tea has about 0.3 mg/ml of caffeine and 0.04 mg/ml of theophylline; these levels are well within the range that should affect amoebae. You may find, as J. T. Bonner did, that as little as 0.025 mg/ml of theophylline can delay aggregation. How might you explain this at the biochemical level? What do you think the effect of caffeine or theophylline would be on a migrating slug or Mexican hat? If you use coffee or tea to ask these questions, I suggest measuring out instant coffee or tea, diluting it with sterile distilled water, recording the concentrations you use, and measuring its pH before applying it.

Behavioral studies

Prior to starvation, *Dictyostelium* amoebae show chemotactic behavior toward the folic acid (one of the B vitamins) that bacteria secrete. You could test the various aspects of this behavior. What concentration of folic acid is effective? (A range between 10^{-6} M and 10^{-2} M are ballpark concentrations to try: folic acid can be bought as a vitamin supplement, a 400 mcg pill dissolved in 10 ml distilled H_2O makes approximately a 10^{-4} *M* solution.) Are cells of a disaggregated slug attracted to folic acid?

Dictyostelium slugs show a number of behaviors. For example, they migrate toward light, higher temperature, higher humidity, low ionic concentrations, and the acidic side of a pH gradient. The cue for settling down into a Mexican hat may be the breaking of the surface film of water by the anterior tip of the slug, which it would do under normal circumstances if it were moving upward, toward light. The presence of ammonia, which is copiously produced by the migrating slug, inhibits culmination. You can design experiments that will test various parameters of behavior. For example, will slugs settle into Mexican hats in the absence of light? Are they more sensitive to certain wavelengths of light than others? If you expose a slug to ammonia will you prevent culmination? And if so, for how long? (10–100 m*M* ammonium hydroxide titrated to pH 7.5 with concentrated phosphoric acid is a concentration range you could use; household ammonia used for cleaning is approximately 3–4 *M*—diluting it to m*M* concentrations is something you could try.) What salts encircling an aggregate or slug prevent migration, and why do you think migration is inhibited? (It is presently not known why.) A number of questions can be asked. The trick is to design a setup that will do the asking. A box with a small hole at one end provides a good chamber for testing movement toward light, and different colors of cellophane can be used over the hole to test different wavelengths of light. You will have to observe your cultures over several days to collect your data.

Dictyostelium fruiting bodies are surprisingly reactive. They move in response to air currents, for example. And if a dissecting needle is placed near the spore mass, the fruiting body may bend in response to the needle. You can experiment with a number of stimuli (sound, air currents, heat) or substances of different composition (wood, plastic, different metals) to try to determine what specifically the fruiting body responds to. You can also try to determine the stimuli that lead to the release of spores. Is it a response to physical touch, to water? Since these aspects of *Dictyostelium* biology have been largely ignored by researchers, your results can add new information to our knowledge of this organism.

Completing your experiments

Watch your cultures carefully throughout the rest of the laboratory period—a lot can happen in only a few hours. Also, check your cultures throughout the week as time allows. Keep careful records of any changes. Always use diagrams to document what you see. A great deal of data will be tied up in these diagrams, so make them accurate and proportional. You may run your experiments as long as you wish, or your energy allows. You do not have to terminate them before the next laboratory period. The only constraint is that they be completed before a report on them is due. Have fun and be imaginative.

Accompanying Materials

Tyler, M. S. and R. N. Kozlowski. 2000. *Vade Mecum: An Interactive Guide to Developmental Biology*. Sinauer Associates, Sunderland, MA. "Slime Mold." This chapter of the CD shows, in movies, time-lapse photography, and still pictures, each of the life stages of *Dictyostelium* and techniques used to set up cultures of the organism.

Gilbert, S. F. 2000. *Developmental Biology*, 6th Ed. Sinauer Associates, Sunderland, MA, Chapter 2. Within this chapter is an excellent, concise summary of the life cycle, biochemistry, and genetics of *Dictyostelium*.

Fink, R. (ed.). 1991. *A Dozen Eggs: Time-Lapse Microscopy of Normal Development*. Sinauer Associates, Sunderland, MA. Sequence 2 is an excellent abridged version of one of John Tyler Bonner's films on *Dictyostelium*.

Selected Bibliography

Berks, M. and R. R. Kay. 1990. Combinatorial control of differentiation by cAMP and DIF-1 during development of *Dictyostelium discoideum*. *Development* 110: 977–984. The authors show that differentiation-inducing factor (DIF) is crucial to the differentiation of stalk cells by activating stalk-specific genes and inhibiting spore-specific genes.

Bonner, J. T. 1957. *Fruiting in* Dictyostelium. Precision Film Labs. This is an 11-minute film by John Tyler Bonner. It is superb, and you should find it tucked away in a film rental library at many Universities.

Bonner, J. T. 1959. Differentiation in social amoebae. *Sci. Am.* 201(6): 152–162. Though old, this beautifully illustrated article is far from outdated. It is a wonderful source of ideas for simple but elegant experiments that could easily be done for this laboratory.

Bonner, J. T. 1983. Chemical signals of social amoebae. *Sci. Am.*, 248 (4): 114–120. A very readable account of sorting out behavior among two species of slime molds.

Bonner, J. T. 1991. *Researches on Cellular Slime Molds: Selected Papers of J. T. Bonner*. Indian Academy of Sciences, Bangalore, India. Bonner's work on *Dictyostelium discoideum* spans five decades and constitutes a major body of information on the organism. His papers are clear and easily understood, with extremely instructive illustrations. The papers are a treasure trove of observations and ideas that will serve students very well as a model for their own reports.

Bonner, J. T. 1998. A way of following individual cells in the migrating slugs of *Dictyostelium discoideum*. *Proc. Natl. Acad. Sci, USA* 95: 9355–9359. This is yet another example of Bonner's genius in devising simple ways of getting an organism to reveal details about its biology. Here, Bonner explains how he viewed individual cell migration by creating thin slugs caught between a coverslip and mineral oil on a slide.

Bonner, J. T., D. S. Barkley, E. M. Hall, T. M. Konijn, J. W. Mason, B. O'Keefe III and P. B. Wolfe. 1969. Acrasin, acrasinase, and the sensitivity to acrasin in *Dictyostelium discoideum*. *Dev. Biol.* 20: 72–87. Acrasin is the old term for what is now known to be cAMP, secreted by aggregating amoebae; acrasinase is phosphodiesterase. This paper includes a smorgasbord of experiments, including ones using human urine and theophylline. (Urine acts as an attractant during aggregation, and theophylline inhibits or delays aggregation.)

Bonner, J. T., T. A. Davidowski, W.-H. Hsu, D. A. Lapeyrolerie and H. L. B. Suthers. 1982. The role of surface water and light on differentiation in the cellular slime molds. *Differentiation* 21: 123–126. A paper that beautifully illustrates how, with a minimum of equipment and astute observation, significant discoveries are still to be made.

Cavalier-Smith, T. 1998. A revised six kingdom system of life. *Biol. Rev.* 73: 203–266. How to classify the cellular slime molds has long been in question.

It is now generally agreed that they belong among the protozoa, and Cavalier-Smith has designated the group as the Mycetozoa.

Devreotes, P. 1989. *Dictyostelium discoideum*: A model system for cell–cell interactions in development. *Science* 245: 1054–1058. A sophisticated paper that discusses primarily the mechanisms of signal transduction through cAMP.

Gerisch, G. and D. Malchow. 1976. Cyclic AMP receptors and the control of cell aggregation in *Dictyostelium*. *Adv. Cyclic Nucleotide Res.* 7: 49–68. Published at a time when cAMP was an extremely hot research topic, this paper summarizes much of the work on cAMP in *Dictyostelium* and provides a wealth of ideas for laboratory studies.

Gomer, R. H. and R. A. Firtel. 1987. Cell-autonomous determination of cell-type choice in *Dictyostelium* development by cell-cycle phase. *Science* 237: 758–762. Gomer and Firtel put forward the theory that the determination of amoebae to prestalk or prespore is controlled by their position in the cell cycle at the onset of starvation.

Gross, J. D. 1994. Developmental decisions in *Dictyostelium discoideum*. *Microbiological Reviews* 58: 330–351. This is a superb summary of much of what we know about the biology and biochemistry of *Dictyostelium*.

Hall, A. L., J. Franke, M. Faure and R. H. Kessin. 1993. The role of the cyclic nucleotide phosphodiesterase of *Dictyostelium discoideum* during growth, aggregation, and morphogenesis: Overexpression and localization studies with the separate promoters of the *pde. Dev. Biol.* 157: 73–84. The complex story of phosphodiesterase in *Dictyostelium* has been unraveled with exquisite finesse by Richard Kessin and his colleagues. This paper summarizes much of that work and demonstrates the delicate precision with which genetics can be studied in this organism.

Kay, R. R. and R. H. Insall, 1994. *Dictyostelium discoideum*. In *Embryos, Color Atlas of Development*. J. B. L. Bard (ed.). Wolfe Publ., London, pp. 23–35. A clear review of the life cycle of *Dictyostelium* with exquisite photographs.

Kessin, R. H. and M. M. Van Lookeren Campagne. 1992. The development of a social amoeba. *Am. Sci.* 80(6): 556–565. If you read no other papers on *Dictyostelium*, read this one. It is a delightfully written, beautifully illustrated article on the general biology, biochemistry, and genetics of *Dictyostelium discoideum*.

Parent, C. A. and P. N. Devreotes. 1996. Molecular genetics of signal transduction in *Dictyostelium. Annu. Rev. Biochem.* 65: 411–440. A clarifying review of the complex of transduction pathways that are part of *Dictyostelium*'s biology.

Powell-Coffman, J. A. and R. A. Firtel. 1993. Cellular dedifferentiation and spore germination in *Dictyostelium* may utilize similar regulatory pathways. *BioEssays* 15: 131–133. This short review discusses the mechanisms of dedifferentiation—how a pseudoplasmodium can be caused to revert back to the stage of feeding amoebae.

Raper, K. B. 1935. *Dictyostelium discoideum*, a new species of slime mold from decaying forest leaves. *J. Agric. Res.* 50: 135–147. This is the report of the discovery of *Dictyostelium* by Kenneth Raper, then a graduate student, who isolated the organism from soil samples taken on a camping trip while in the Smoky Mountains of North Carolina.

Riley, B. B. and S. L. Barclay. 1990. Ammonia promotes accumulation of intracellular cAMP in differentiating amoebae of *Dictyostelium discoideum*. *Development* 109: 715–722. The paper describes experiments in which ammonia and high pH were used to promote spore and inhibit stalk formation. The ammonia caused an increase in intracellular cAMP by preventing its secretion.

Extracellular cAMP levels were thereby lowered.

Riley, B. B. and S. L. Barclay. 1990. Conditions that alter intracellular cAMP levels affect expression of cAMP phosphodiesterase gene in *Dictyostelium*. *Proc. Natl. Acad. Sci. USA* 87: 4746–4750. This paper discusses the effects of caffeine on *Dictyostelium*, showing that the resulting increase in phosphodiesterase activity leads to increased stalk and decreased spore differentiation.

Shaulsky, G., D. Fuller and W. F. Loomis. 1998. A cAMP-phosphodiesterase controls PKA-dependent differentiation. *Development* 125: 691–699. An important paper on the control points of transduction pathways involving cAMP in *Dictyostelium*.

Suppliers

Carolina Biological Supply Co.
2700 York Road
Burlington, NC 27215
1-800-334-5551
www.carosci.com

Connecticut Valley Biological Supply Co., Inc.
P.O. Box 326
82 Valley Road
Southampton, MA 01073
1-800-628-7748

Dictyostelium discoideum. Four or five tubes will seed up to 50 petri dishes. Order two to three weeks in advance. If ordering in the winter, ask for cultures with mature stages of *Dictyostelium*. Spores can survive an accidental freeze, but earlier stages can not.
Escherichia coli (non-mucoid) One or two tubes can be used to grow a lawn of bacteria on nutrient agar, enough for an entire class.

Any good biological supply company, such as:
Fisher Scientific
585 Alpha Dr.
Pittsburgh, PA 15238
1-800-766-7000
www.fishersci.com

0.22-μm porosity sterilizing filters and filtering apparatus
Bacteriological loops
Nutrient agar supplies
Petri dishes
Vital dyes

Any good chemical supplier, such as:
Sigma Chemical Company
P.O. Box 14508
St. Louis, MO 63178
1-800-325-3010
www.sigma-aldrich.com

Adenosine 3´, 5´-cyclic monophosphate (cAMP)
Caffeine
Theophylline

c h a p t e r 5 *Gametogenesis*

It is typical in general biology courses to learn about mitosis and meiosis all in one breath. This is unfortunate, for it dulls the fascination for the unique nature of meiosis. Think about meiosis: Where does it occur? Only in the gonads—the ovary and testis—and more specifically, only in the germ cells that are lodged in the gonads. Meiosis is the method of cell division that halves the number of chromosomes in the resulting egg and sperm cells. These cells, the gametes, can then fuse with gametes of the opposite type (an egg with a sperm) without multiplying the total number of chromosomes in the resulting zygote. Instead, the normal diploid number of chromosomes is restored.

Meiosis is an exotic and profound event. It is the foundation for sexual reproduction. Without meiosis, there would be no sex. Ponder this, and think about its significance as you study the organs that specialize in this event.

Meiosis: An Outline

Primordial germ cells arrive in the developing gonad after a long trip, often from the far reaches of the yolk endoderm. Their first task is to increase in number, which they achieve by standard mitosis. This is the **gonial stage**. Developing eggs in this stage are called **oogonia** and developing sperm, **spermatogonia** (Figure 5.1). Germ cells leave the mitotic stage to enter a stage of growth—the **primary gonocyte stage**—that includes **primary oocytes** and **primary spermatocytes**. Primary gonocytes then enter the first meiotic division. During this division, homologous chromosomes separate. The daughter cells are haploid, and each of their chromosomes is made up of two chromatids. This is the **secondary gonocyte stage**, which includes **secondary oocytes** and **secondary spermatocytes**. These enter the second meiotic division, during which the chromatids of each chromosome separate from one another. The resulting cells, **gonotids** (**ootids** and **spermatids**), are still haploid.

As described above, every primary gonocyte has the potential to produce four gonotids. This is what occurs in developing sperm. The meiotic divisions are equal, and each primary spermatocyte produces four equal size, small spermatids. The spermatids then go through a further differentiation process called **spermiogenesis** to become fully formed spermatozoa.

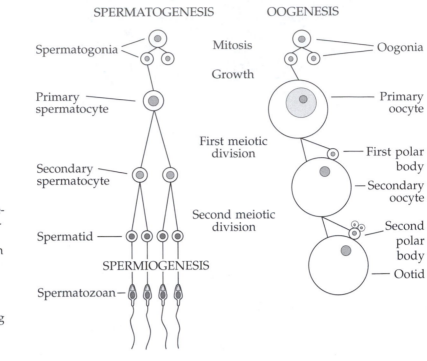

SPERMATOGENESIS   OOGENESIS

Spermatogonia   Mitosis   Oogonia

Growth

Primary spermatocyte   Primary oocyte

First meiotic division

Secondary spermatocyte   First polar body

Secondary oocyte

Second meiotic division

Spermatid   Second polar body

SPERMIOGENESIS   Ootid

Spermatozoan

Figure 5.1
Schematic diagram of gametogenesis. Mitosis in the gonial stage is followed by a period of growth. Cells then enter meiosis. The meiotic divisions in developing sperm are equal. In the developing ovum, however, they are unequal, producing polar bodies and a single large ovum.

Meiotic divisions in eggs, on the other hand, do not produce four equal size cells. Instead, on each meiotic division a very eccentric spindle produces one large cell and one tiny cell. A single primary oocyte, therefore, produces a single large cell, the **ootid**, and two tiny cells, the **polar bodies**. (Sometimes the first polar body divides along with the oocyte on the second meiotic division, making a total of three polar bodies.) The polar bodies sit in a clump like bumps on the large spherical ootid and are eventually discarded. Egg development, therefore, involves a lot of genetic waste. For every egg that is produced, three potential egg cells are wasted, all for the sake of retaining as much cytoplasm as possible in the single ootid. Why do you think this is important or necessary? Record your answer in your laboratory notebook.

You will be using slide material to study this world of reduction divisions and genetic waste. We will concentrate on mammalian gametogenesis because of its inherent interest to us as gamete-producing mammals.

You will be given sections of mammalian (probably rat) testis and ovary on slides. The sections are thin, approximately 6–10 µm, or the thickness of a cell. Think about how thick this is. Scratch your head for a flake of dandruff. The flake is a single epithelial cell from your scalp. It too is about 10 µm thick. Now look back at your slides. The sections are not the color of dandruff, but are an array of at least two contrasting colors. Check the label of the slides to see if you can determine which stains were used. The most commonly used histological stains are hematoxylin and eosin. Hematoxylin is blue-to-black and has an affinity primarily for nucleic acids. The nuclei, therefore, will be stained blue-black. Eosin, a red color, stains the cytoplasm. The contrast of blue against red facilitates the definition of individual cells.

Place both slides side by side against a white background (such as a note card), and look at them under the dissecting scope. Note as many features as you can about both the testis and the ovary before proceeding with the detailed instructions. Use your micrometer to measure the dimensions of the ovary and testis. Which one is bigger? Record your measurements in your laboratory notebook.

Mammalian Spermatogenesis

As you look at the testis slide under the dissecting scope, you should see a roundish structure surrounded by a tough sheath of connective tissue. This is the **tunica albuginea**. Within the roundish structure are numerous cross sections through tubules, the **seminiferous tubules** (Figure 5.2). These tubules should look like little doughnuts. It is in these seminiferous tubules that the sperm are being produced. Between the tubules, you should see small, triangular islands of tissue containing blood vessels and the **interstitial cells**, or **Leydig cells**, which secrete the male sex hormone, testosterone.

Most males are incredible sperm-producing machines. The human male, for example, can produce up to 300 million sperm per day and can pack the same amount into a single 3-ml ejaculate. Any organ that can mass-produce a cell to this extent must have an enormous surface area for the task. And this is the case. The seminiferous tubules, U-shaped structures each about 80 cm long, are packed into 250 testicular lobules, each with 1–4 tubules. If all the seminiferous tubules from just one testis were strung end to end, they would extend about 255 meters (or 279 yards). How far is this? Go outside to the track and run as fast as you can for 30 seconds. Now measure off how far you have run. Unless you are quite extraordinary, it will be about halfway around, or about 220 yards, less than the length of the seminiferous tubules from one testis.

This extensive sperm-producing capacity might seem like overkill. However, quantity appears to have its purpose. In humans, a sperm count of less than 20 million sperm per ml of ejaculate is tantamount to being sterile. Apparently, the huge numbers are necessary in helping the one successful sperm to its mark.

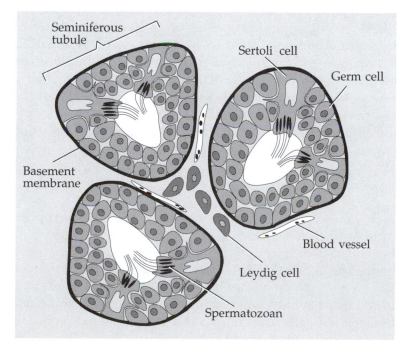

Figure 5.2
Diagram of a cross-section through seminiferous tubules within a mammalian testis. Three seminiferous tubules are shown containing developing sperm and Sertoli cells. The interstitial region between the tubules contains Leydig cells and blood vessels.

Other causes of sterility in men can be tight jeans, hot baths, and fevers, all of which raise the temperature of the testis. Sperm development is extremely temperature-sensitive and can only occur in mammals at a temperature 1.5–2.0°C cooler than normal body temperature. This is why the testes are held in scrotal sacs outside the body cavity. In humans, sperm production starts declining at approximately age 45, but often not enough to cause sterility, as is evidenced by the many men who have fathered children in their later years.

Stereology

Now examine the section under the low power of your compound microscope. Use your micrometer to estimate the thickness of the tunica albuginea and the diameter of a seminiferous tubule. Record these in your laboratory notebook. You have already determined the dimensions of the section. Now determine the area of your section of testis (since the section is roughly circular, you can determine area using $A = \pi r^2$, where $\pi = 3.14$). Use your micrometer grid to determine how many cross sections of seminiferous tubules occur in 1 square millimeter. Now, knowing the area of your section, determine how many cross sections of seminiferous tubules occur in the entire section. Since each tubule is U-shaped, half this number should give an approximation of the number of tubules within the testis. If each one were 80 cm long, and they were strung end to end, how far would they extend? How does this number compare to that in humans?

You can determine the surface density of the seminiferous epithelium by making a simple calculation used in stereology. First, choose one set of lines (horizontal or vertical) on your micrometer grid and figure out their total combined length. Then count the total number of intercepts these lines make with the tubules. (Any tubule lying on a line intercepts it twice.) This number, multiplied by 2 and divided by the combined length of the lines (in µm) equals the number of square micrometers of epithelium in a cubic micrometer of tissue. Now, if you know the volume of the testis, you can calculate the total area of seminiferous epithelium within it (since the testis is roughly spherical, you can estimate volume using $V = 4/3 \, \pi r^3$). The number will be impressive.

Tunica albuginea

Look at the very thick **tunica albuginea**. What color does it stain? This will tell you what color collagen stains in this preparation, since the tunica albuginea consists primarily of extracellular fibers of type I collagen. The cells that make this collagen, wherever it occurs in the body, are called **fibroblasts**. You will see them as small, spindle-shaped cells, embedded among the layers of collagen.

Use polarized light to observe the tunica albuginea. Does it shine under crossed polarizers? If so, or if not, what does this tell you about the arrangement of collagen fibers here?

Seminiferous tubules

The **seminiferous tubule** is the epithelium that produces sperm. Being an epithelium, it has a free apical surface facing the lumen of the tubule, and a basal surface that sits on a **basement membrane**. The basement membrane for any epithelium is of major importance to the differentiation and maintenance of that epithelium. It is an acellular layer, made up primarily of type IV collagen, laminin, fibronectin, and heparan sulfate proteoglycan. Examine a seminiferous tubule with zbasement membrane. It will be very thin. What color does it stain? Now look back at the tunica albuginea. What color does it stain? Since the major component in each of these is collagen, both should stain the same color. Examine the basement membrane under polarized light. Is it different from the tunica albuginea? What does this tell you?

Look at the epithelium carefully. It is made up of two types of cells, the male **germ cells** at various stages of differentiation as they become spermatozoa, and nurse cells called **Sertoli cells**. Since development of the germ cells is synchronous at any one level within the tubule, but out of synchrony with other levels, you will see different stages of development in different tubules. By looking at several tubules, you will be able to see germ cells in the spermatogonial, primary spermatocyte, secondary spermatocyte, spermatid, and spermatozoan stages (Figure 5.3). As the cells progress through their differentiation, they move closer and closer to the lumen of the tubule. Spermatozoa sit with their tails sticking out into the lumen.

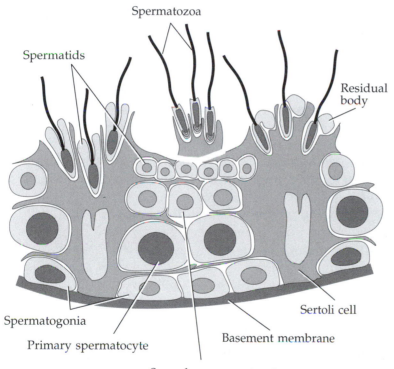

Figure 5.3
Schematic diagram of a section through a mammalian seminiferous tubule. All stages of sperm development are shown. Spermatogonia are undergoing mitosis. The primary spermatocytes are in a period of growth. In leaving this stage, they divide meiotically to produce secondary spermatocytes, which quickly divide in a second meiotic division to produce spermatids. The spermatids differentiate into spermatozoa, a process called spermiogenesis. Three different stages in this process are shown (although all three stages would never be seen together as drawn): an early stage where the nucleus is condensing and the tail is forming; a later stage where excess cytoplasm is being pinched off as a residual body; and a final stage where the spermatozoa are ready for release. Since sperm development is synchronized at any one level within a tubule, you must look at different tubules or at different levels along a single tubule to see the stages represented.

The **Sertoli cells** are nondividing, tall, columnar cells that extend from the basement membrane to the lumen of the tubule. The nuclei of these cells are irregularly shaped and stain lightly, making them easy to identify. Though one cannot tell from light-microscopic study, the Sertoli cells are contiguous, making an unbroken ring around the tubule. The germ cells at all stages of development sit nested in indentations of the Sertoli cells, a fact implying that the Sertoli cells function to support and nourish the germ cells throughout their development. Sertoli cells are also responsible for the translocation of the developing sperm toward the lumen and their eventual release. One of their truly significant functions is to create the blood-testis barrier. The developing sperm are isolated from any immune response as long as they are within the seminiferous tubules, since no immunoglobulins can pass this blood-testis barrier. Why do you think such a thing is necessary? (Remember that the haploid sperm are genetically different from the diploid cells of the body.) If a man has a vasectomy (a tying off of the ductus deferens, the duct leading from the epididymis, as a method of contraception), an immune response is launched against sperm that sit trapped in the ducts above the level of the ligature. If the man has the vasectomy surgically reversed in order to father children, the built-up immune response against his sperm can lead to sterility.

Among the Sertoli cells will be the numerous smaller, more darkly stained nuclei of the germ cells. The **spermatogonia**, with their dark, granular, round or slightly oval nuclei, sit against the basement membrane. They are multiplying through mitosis, so you may see mitotic figures in this layer. The **primary spermatocytes** sit as a layer above the spermatogonia. Remember that they are in a period of growth, so they are the largest germ cells. Their nuclei also are large and stain relatively darkly. The first meiotic division gives rise to **secondary spermatocytes**, but you will be lucky to see any, since they almost immediately go into the second meiotic division. You can find them by looking for cells that are half the size of primary spermatocytes. Remember, primary spermatocytes are still diploid, and secondary spermatocytes are haploid.

The second meiotic division produces **spermatids**. These are one-quarter the size of primary spermatocytes and are closest to the lumen. Try to find meiotic figures. You can distinguish first maturation division figures as being larger than second maturation division figures. Diagram the stages of spermatogenesis as seen in tubules on your slide, labeling all structures that are in boldface type above.

Spermiogenesis

The spermatids in some tubules form a layer three or four cells thick. They are going through a complex process of differentiation called spermiogenesis to become spermatozoa. You should be able to see the following parts of the process on your slide:

1. Elongation of the nucleus and cytoplasm of the spermatid.
2. Condensation (shrinkage and compaction) of the nucleus in what is becoming the head of the sperm and formation of an **acrosomal cap**.
3. Formation of a **flagellum** that extends toward the lumen of the tubule.
4. An extensive shedding of cytoplasm that pinches off and appears as dark **residual bodies** near the lumen of the tubule. These residual bodies will be phagocytosed (eaten) by the Sertoli cells.

When the sperm are fully formed spermatozoa, only the tip of the head remains attached to a Sertoli cell. See if you can find any fully formed spermatozoa. Eventually, the Sertoli cell will release the spermatozoa into the fluid of the lumen. The entire process of spermatogenesis (from spermatogonium to spermatozoan) takes 48 days in the rat and 64 days in humans. Draw each of the stages of spermiogenesis that you are able to find in your section.

Interstitial cells

The **interstitial cells**, or **Leydig cells**, lying between the seminiferous tubules constitute the mesenchyme of the testis. Look at these cells again. They are large, about 20 μm in diameter, with a spherical, lightly staining nucleus. Though they are relatively few in number, their role is extremely important. In response to leutinizing hormone, LH (yes, I know, you thought only females produced LH), the Leydig cells secrete testosterone, which is required both for spermatogenesis and for development of secondary male characteristics. (Follicle stimulating hormone, FSH, is also produced in males. Its target is the Sertoli cells, causing them to increase their cAMP levels and secrete androgen binding protein.)

Comparative Sperm Morphology

Obtain a slide of guinea pig sperm. These are extremely large sperm. Examine a sperm under high power. The **acrosome** should be obvious (Figure 5.4). This is the bag of enzymes that will open during fertilization in the acrosome reaction. The released enzymes coat the head of the sperm and break down the various extracellular layers surrounding the egg, allowing the sperm smooth sailing toward the egg cell surface. Try to focus on the midpiece and tail of the sperm. Remember that the flagellar tail is near the limits of resolution of your microscope. Use the methods that give you the best resolution of your microscope (Koehler illumination,

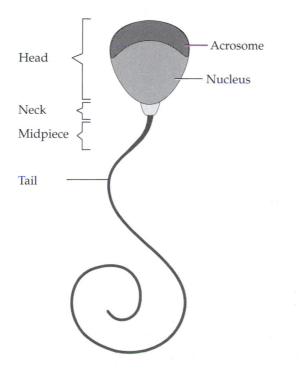

Figure 5.4
Diagram of a guinea pig spermatozoan. The acrosome is a bag of enzymes used during fertilization. The nucleus is highly condensed.

blue filter; look back at your microscopy lab). Now use various techniques to increase contrast. Which of the methods gives you the best results? Record this information in your laboratory notebook, and make a labeled diagram of the guinea pig sperm.

Examine the various types of sperm that are on display. Notice the variety of sizes and shapes that exist among species. Make a diagram of each type that you see, being sure to indicate its species.

Mammalian Oogenesis

The ovary does not have the capacity for germ cell production that the testis does. A woman is born with only about 800,000 eggs, and that's all she'll have for her lifetime. The oogonial stage, when germ cells are increasing by mitosis, occurs prior to birth, and by the time of birth all the germ cells exist as primary oocytes. During the 30–40 years that a woman has menstrual cycles, she will ovulate about 400 eggs; the remaining 99+% of eggs undergo atresia, or degeneration. (Though this may seem like a tremendous waste, it doesn't come close to the level of unused sperm males produce.)

Look at your slide of mammalian (probably rat) ovary against a white background under the dissecting scope. You will see a thin connective-tissue sheath, the **tunica albuginea**, surrounding the teardrop-shaped ovarian tissue. Notice how thin this tunica albuginea is compared to that surrounding the testis. On the narrow end of the ovary, a region called the **hilum**, you may see the remnants of the **mesovarium**, a narrow rope of connective tissue that attached the ovary to the broad ligament within the body cavity.

Within the ovary you should see two regions: an outer **cortex**, containing circular structures called **follicles**, and an inner **medulla**, containing connective tissue, nerves, lymph vessels, and many large blood vessels. The demarcation between these two regions is indistinct. The germ cells are all located in the follicles within the cortex. A follicle consists of a single egg surrounded by one or more layers of nurse, or follicle, cells.

Now look at your slide under the low power of your compound microscope. Look again at the tunica albuginea. You might see a single layer of small, cuboidal cells just outside of the tunica albuginea. This is the **mesothelium**, an epithelium that is often accidentally brushed off during histological processing.

Make a diagram of the ovary, and add to it the details of the structures as they are covered in the instructions below.

Follicles

Under low power, and then higher power, examine the follicles (Figures 5.5 and 5.6). **Primordial follicles** consist of a primary oocyte, arrested in its first meiotic division, surrounded by a single layer of squamous follicle cells. You should see many of these. Most of the follicles are at this stage. Starting at puberty, and continuing until menopause, each menstrual cycle begins with the growth of 20–50 primordial follicles, stimulated by a pituitary hormone, follicle stimulating hormone (FSH). As soon as the follicles begin to grow, they are known as **primary follicles**. First, the follicle cells become cuboidal and/or columnar, and they increase in numbers. The multiple layers of follicle cells are the **stratum granulosum** (these cells will be providing nutrients for the egg). The follicle cells are also secreting a

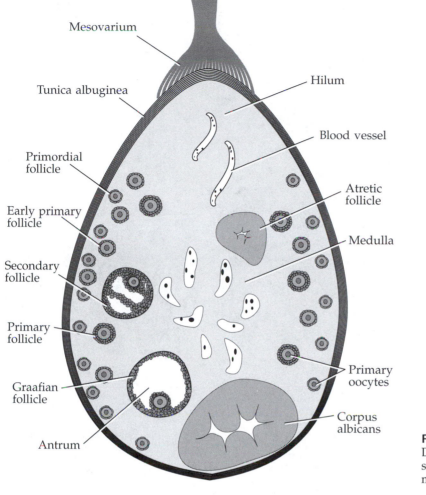

Mesovarium

Tunica albuginea

Hilum

Blood vessel

Primordial
follicle

Atretic
follicle

Early primary
follicle

Medulla

Secondary
follicle

Primary
follicle

Primary
oocytes

Graafian
follicle

Corpus
albicans

Antrum

Figure 5.5
Diagram of a longitudinal
section through a mam-
malian ovary.

fluid, the **liquor folliculi**, which gradually accumulates in spaces that develop between follicle cells. Once these spaces appear, the structure can be called a **secondary follicle**. The spaces constitute the **antrum**, which eventually becomes
enormous, filling the entire central region of the follicle. The egg at this point sits
in a little hummock of follicle cells called the **cumulus oophorus**. Such a follicle is
now mature, and can be called a **tertiary** or **Graafian follicle**. (It is named after its
discoverer, the Dutch anatomist Regnier de Graaf, who first described the ovarian
follicle in 1672—not the staid, elderly anatomist we might picture, but a young
man of 31. Sadly, he died only a year later.) It is from a Graafian follicle that an egg
ovulates.

The Graafian follicle contains large amounts of liquor folliculi. This clear, vis-
cous, nutritive fluid contains high amounts of hyaluronate (a glycosaminoglycan,
or GAG) as well as hypoxanthine and adenosine. The last two substances have
been shown in vitro to keep the oocyte in meiotic arrest, a function they most likely also serve in vivo.

Not all growing follicles get to be Graafian follicles. In humans, for example,
among the 20–50 that start out on the road to maturation each cycle, usually only
one—the **dominant follicle**—becomes a Graafian follicle. The others undergo

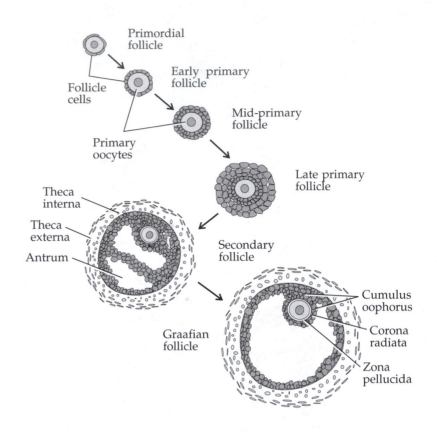

Figure 5.6
Schematic diagram of stages
in follicle development.

atresia, or degeneration, at various points along the way. These **atretic follicles**, therefore, will be numerous in your section. The cells will look misshapen, and the nuclei will stain darkly. (In mammals with multiple births, such as the rat, more than one Graafian follicle develops each cycle, of course.)

Using your dissecting scope and low power on the compound microscope, measure the diameters of each type of follicle with your micrometer. Record these measurements in your laboratory notebook.

Theca

As the follicles mature, cells outside the follicular cells coalesce around the follicle, forming a **theca**. As the theca becomes more pronounced, two layers can be distinguished, the **theca interna** closest to the follicle, and a more fibrous outer **theca externa**. The theca interna secretes nutritive fluid for the follicle cells, as well as some estrogen- and androgen-intermediary compounds that the follicle cells then convert to estrogen.

Oocyte

Look carefully at the oocytes in each type of follicle. Notice how much larger the oocyte becomes during follicular growth. The oocyte in a primordial follicle is about 25–30 μm in diameter. Using this as a guide, estimate how large the oocyte is in each of the other types of follicles, and record your number. Before it ovulates, the oocyte has grown considerably. This should impress upon you that the primary oocyte stage truly is a stage of growth.

Now focus on an oocyte in a growing follicle. Notice that immediately surrounding the egg is a translucent area. This is the **zona pellucida**. It contains various glycoproteins, named ZP1, ZP2, and ZP3. These play a major role in fertilization, inducing the acrosome reaction in the sperm and eventually being altered by cortical granule substances released by the fertilized egg. In their altered form they prevent polyspermy.

Notice that traversing the zona pellucida are many fine, filamentous structures. Use various methods to increase contrast in your microscope in order to see these better. These are **cytoplasmic bridges** between the follicle cells immediately surrounding the oocyte and the oocyte. Nutritive substances such as yolk proteins are being pumped through these channels into the egg during the growth phase.

This layer of follicle cells immediately surrounding the egg is called the **corona radiata** (meaning "radiating crown"). When the oocyte is ovulated, this layer will remain attached to the egg and travel with it into the oviduct. At fertilization, the sperm will have to penetrate the corona radiata and the zona pellucida in order to get to the egg surface.

Ovulation of the oocyte is triggered by a surge in luteinizing hormone secreted by the pituitary. Just prior to ovulation, the oocyte completes the first meiotic division and is ovulated as a secondary oocyte. The second meiotic division will begin in the upper end of the oviduct, only to once again enter meiotic arrest. It will not complete meiosis until after fertilization. It is unlikely that you will be fortunate enough to see a secondary oocyte. Look for an oocyte within a Graafian follicle that has a small **first polar body** on its surface.

Corpus luteum

Once the oocyte is ovulated, the Graafian follicle becomes an endocrine organ, secreting progesterone and estrogen, which maintain the uterine lining in preparation for implantation. The follicle is now the **corpus luteum** (meaning "yellow body"). It looks collapsed, and its walls are thrown into folds. This structure is huge—1.5 to 2 cm in diameter. Find a corpus luteum on a slide marked as having one. Under a dissecting scope, use your micrometer to measure it, and make a record of these measurements.

If pregnancy does not occur, the corpus luteum declines. In humans, this occurs after 14 days, and its decline brings on menses. The declining corpus luteum is replaced by scar tissue and becomes a smaller **corpus albicans** (meaning "white body"). Find a corpus albicans, measure its dimensions, and make a record of these. How much smaller is it than the corpus luteum?

If pregnancy does occur, then the corpus luteum grows. In humans it reaches as much as 5 cm in diameter. (This is really big!) The corpus luteum of pregnancy continues to secrete hormones until the ninth or tenth week of pregnancy, when the placenta takes over this job. This enormous corpus luteum then involutes, and is replaced by scar tissue, becoming a very large corpus albicans. The scar is so pronounced that it usually causes retraction of the ovarian surface. Try to find one of these in your section. Make a record of its measurements.

Histological Hitchcocking

It is now time to test your sleuthing abilities. If you have learned your histology well, you will be as clever as the judge below.

Little weeping and wailing was heard when the reclusive widow died. She was rich and had kept all her money in nylon stockings stashed between the mattress and box springs. The lawyer assigned to the case found that there was no will; however, two people had already stepped forward claiming rights to the inheritance. The first was a man who claimed he was the widow's only child and should inherit all. The second was a woman claiming to be this man's sister, who insisted that the inheritance be shared. The case went to court.

The judge listened to the tads of evidence and made a careful study of the pathologist's report from the autopsy. This judge had been a stellar student in developmental biology, and she knew her stuff. Based solely on information in the pathologist's report, she determined that the woman was lying and awarded the entire inheritance to the man.

How did the judge know? What was that piece of evidence she found in the pathologist's report? Record your answer in your laboratory notebook.

Accompanying Materials

Tyler, M. S. and R. N. Kozlowski. 2000. *Vade Mecum: An Interactive Guide to Developmental Biology*. Sinauer Associates, Sunderland, MA. "Gametogenesis." This chapter of the CD explains meiosis and illustrates in histological sections the process of gametogenesis in mammalian gonads. It also steps you through a stereology exercise using the testis to determine the amount of surface area devoted to sperm production.

Gilbert, S. F. 2000. *Developmental Biology*, 6th Ed. Sinauer Associates, Sunderland, MA. Chapters 7 and 19. These two chapters discuss meiosis and the structure of gametes.

Fink, R. (ed.). 1991. *A Dozen Eggs: Time-Lapse Microscopy of Normal Development*. Sinauer Associates, Sunderland, MA. Sequence 4 shows meiosis in a flatworm egg, along with the impressive surface blebbing that occurs prior to each division.

Selected Bibliography

Cobb, J. and M. A. Handel. 1998. Dynamics of meiotic prophase I during spermatogenesis: from pairing to division. *Sem. Cell. Dev. Biol.* 9: 445–450. A review of meiosis in mammalian spermatogenesis.

Cormack, D. H. 1987. *Ham's Histology*. 9th Ed. J. B. Lippincott, Philadelphia. This is a superb text that includes a lot of incidental information along with the straight histological descriptions.

Elias, H. and D. Hyde. 1983. *A Guide to Practical Stereology*. S. Karger Publ., New York. Any histologist vastly expands the data he or she can squeeze out of a slide by investing a little time in stereology. This short manual is user-friendly, written for the biologist, and will prevent panic attacks even in a severe mathophobe.

Fawcett, D. W. 1994. *A Textbook of Histology*, 12th Ed. Chapman and Hall, N.Y. Originally published with Bloom, this tome is one of the few left from the era when the cell biology, ultrastructure, and light-microscopic details of tissues were taught in a single course. It is superb in its coverage; the extent of details on the mammalian gonads is without match.

Frayne, J., and L. Hall. 1999. Mammalian sperm-egg recognition: does fertilin β have a major role to play? *BioEssays* 21: 183–187. This explains, with colored diagrams, the role of components of the zona pellucida and the sperm cell membrane in the fertilization reaction.

Hecht, N. 1998. Molecular mechanisms of male germ cell differentiation. *BioEssays* 20: 555–561. This review gives a detailed account of the nuclear events during mammalian spermatogenesis.

Joyce, I. M., F. L. Pendola, K. Wigglesworth and J. J. Eppig. 1999. Ooctye regulation of kit ligand expression in mouse ovarian follicles. *Dev. Biol.*, 214: 342–353. A research article that demonstrates that mammalian oocytes influence gene expression in their surrounding follicle cells. It's an excellent example of current research on mammalian oogenesis.

Moore, K. L. and T. V. Persaud. 1998. *Before We Are Born*. 5th Ed. W. B. Saunders, Philadelphia. The second chapter of this very readable short text is an excellent summary of human reproduction.

Picton, H., D. Briggs and R. Gosden. 1998. The molecular basis of oocyte growth and development. *Molecular and Cellular Endocrinology* 145: 27–37. An excellent review of mammalian oogenesis.

Reider, C. L. (ed.). 1999. *Methods in Cell Biology, Vol. 61, Mitosis and Meiosis*. Academic Press, San Diego. A wealth of valuable information on methods for studying the machinery of cell division.

Ross, M. H., L. J. Romrell and G. I. Kaye. 1995. *Histology: A Text and Atlas*. 3rd Ed. Williams & Wilkins, Baltimore. The illustrations in this text are exquisite. In addition to copious colored diagrams and photomicrographs throughout, there are a series of full-page plates at the end of each chapter.

Suppliers

Prepared microscope slides of testis, sperm, and ovary can be obtained from such suppliers as:

Turtox/Cambosco, Macmillan Science Co., Inc.
8200 South Hoyne Ave.
Chicago, IL 60620
1-800-621-8980

Triarch, Inc.
P.O. Box 98
Ripon, WI 54971
1-800-848-0810

Connecticut Valley Biological Supply Co., Inc.
P.O. Box 326
82 Valley Road
Southampton, MA 01073
1-800-628-7748

Ward's Natural Science Establishment, Inc.
P.O. Box 92912
5100 West Henrietta Road
Rochester, NY 14692-9012
Tel. 1-800-962-2660
www.wardsci.com

chapter 6
Echinoid Fertilization and Development
Sea Urchins and Sand Dollars

Collecting

Have you watched seagulls in flight drop them to smash on the rocks below, pulled the prickly ones from their rocky crevices, or wriggled the fragile, disc-shaped ones from the sand with your toes? Sea urchins and sand dollars, a favored delicacy of seagulls, have also been favored organisms of study for the developmental biologist since the early 1900s. Their development, almost identical, is easily studied since their gametes are readily spawned—and spawned in large numbers. The only problem is finding the adults when they are ripe. If you need gametes in winter, sea urchins are the choice, but they're often hard to get. On the Northeast coast, the threat of winter ice causes sea urchins to migrate out into deep water; they no longer sit conveniently in the intertidal zone to be picked up by the local kids or developmental biologists. They don't return to the intertidal zone until their spawning season is virtually over. The common green sea urchin, for example, is ripe and ready to spawn starting in January. If you don't have a wet suit for diving, you will need accomplices.

One good accomplice is the person who sets lobster traps. Sea urchins get caught in lobster traps and are usually discarded. The lobster pound can also be a good source. If you're willing to compete with the Japanese food trade, you can try one of the many sea urchin processing plants that now dot the coastline. In the past decade, sea urchin processing in this country has escalated into a huge industry that provides the Japanese dinner table with raw ripe gonads. Ships are collecting the urchins for the processing plants just at the time that developmental biologists need them.

If you have no accomplices, you can always obtain sea urchins by the more mundane method of placing an order with a biological supply company. Always know the species being provided, the time that they are ripe, and the temperatures that they tolerate (see Table 6.1).

When transporting and holding ripe urchins, it is best to pack them in wet seaweed or put them individually into small plastic containers of seawater. Keep them cold (4–8°C). If they are held all together in a single tank of seawater, one sea urchin spawning in the tank will stimulate all the others to spawn.

If you don't have access to seawater, it is easy enough to make artificial seawater or to buy instant ocean from a supply company.

Table 6.1 Availability of gametes from common North American sea urchins and sand dollars

Species	North American distribution	Ripe gametes found[a]	Recommended temperature	Egg diameter (μm)
Sea Urchins				
Arbacia punctulata Atlantic purple sea urchin	East coast: Cape Cod to Florida; Texas; Yucatán	June–August	20°C–24°C (above 30°C is lethal)	79
Lytechinus pictus White sea urchin	West coast: Southern California–south	spring–fall	16°C–20°C (below 8°C is lethal)	111
Lytechinus variegatus Variegated urchin	East coast: North Carolina to Florida	fall–spring	22°C–28°C	103
Strongylocentrotus droebachiensis Green sea urchin	East coast: Arctic to New Jersey West coast: Alaska to Puget Sound	January–June	0°C–10°C	160
Strongylocentrotus franciscanus Red sea urchin	West coast: Alaska to Southern California	March–July	10°C–17°C	140
Strongylocentrotus purpuratus Purple sea urchin	West coast: British Columbia to Baja California	December–May	13°C–15°C	80
Sand Dollars				
Dendraster excentricus Eccentric sand dollar	West coast: Alaska to Baja California	Late spring–summer	10°C–22°C	114
Echinarachnius parma Common sand dollar	East Coast: Labrador to Cape Hatteras West coast: Alaska to Puget Sound	summer	4°C–16°C	145

[a]Varies somewhat along its north–south distribution.
Source: Primarily after Czihak, 1975 and Strathmann, 1987.

Artificial seawater

NaCl	28.32 gm	*Dissolve in l liter distilled water.*
KCl	0.77 gm	*Add 0.2 gm NaHCO$_3$*
MgCl$_2$·6H$_2$O	5.41 gm	*Adjust to pH 8.2, if necessary.*
MgSO$_4$·7H$_2$O	7.13 gm	
CaCl$_2$	1.18 gm	

Fertilization

The events of sea urchin fertilization have been worked out in great detail. You will be studying both normal fertilization and parthenogenesis (development without fertilization). You will be able to ask some very sophisticated biochemical questions once you know some of the details of these events.

We will be injecting potassium chloride (KCl) to induce spawning in both male and female sea urchins. KCl causes a contraction of smooth muscles, and immature gametes will be spawned along with ripe gametes. This can cause problems, and you must be able to recognize a mature egg from an immature one. In the sea urchin, eggs are mature when they are ootids. This means that meiosis is completed and the nucleus is relatively small. An immature egg will be a primary oocyte with a huge nucleus (called a **germinal vesicle**). The germinal vesicle will be easily recognizable under a dissecting microscope, if you adjust the light to obtain good contrast and focus up and down through the egg.

The egg has two extracellular coats: a **vitelline envelope**, which before fertilization fits snugly around the egg surface and cannot be distinguished; and an outer layer of jelly. This jelly contains a chemical sperm attractant (a small polypeptide that is species-specific). It is not present in the jelly of an immature egg. From a mature egg's jelly, this attractant diffuses outward, and sperm swim up the concentration gradient. The jelly also contains a relatively species-specific fucose-containing polysaccharide that activates the acrosome reaction in the sperm. This polysaccharide binds to glycoprotein receptors on the head of the sperm, causing the **acrosomal vesicle** within the head to fuse with the cell membrane and release its enzymes. These enzymes coat the head of the sperm and eat through the jelly, making a path to the egg cell surface for the swimming sperm. In the process, the acrosomal vesicle becomes inverted and greatly elongated by the assembly of actin filaments. This elongate structure, the **acrosomal process**, is what will fuse with the egg cell surface, and it can be seen under the compound microscope under oil immersion if the contrast is maximized.

The sperm first binds to the vitelline envelope using a species-specific cell surface protein called **bindin**, which binds to a bindin-receptor protein on the vitelline envelope. The sperm then fuses with the egg cell membrane, and in so doing causes a brief influx of sodium ions. This influx raises the resting membrane potential of the egg from −70 mV to above 0 mV. Sperm can not fuse with an egg whose membrane potential is above about −10 mV, so this change in membrane potential effectively prevents any additional sperm from fusing. This is called the **fast block to polyspermy**. It takes only one-tenth of a second to occur, but is not permanent, lasting only about a minute. Since the fast block depends on the availability of sodium in the medium, you can circumvent it by keeping the sodium concentrations in the surrounding seawater artificially low.

Binding of the sperm with the egg cell membrane also sets up a second block to polyspermy, the **slow block**. This block takes a minute to occur, and it is permanent. You will see it as a lifting of the vitelline envelope away from the egg cell surface. This membrane then toughens (involving a chemical process much like tanning leather) and is now called the **fertilization envelope**. Sperm cannot penetrate this tough layer. The point of sperm entry can be identified by a cone-shaped elevation called the **fertilization cone**. It represents a tangle of microvilli that have elongated and wrapped themselves around the sperm.

The biochemical events that are causing the lifting of the vitelline envelope are initiated through the **phosphatidylinositol bisphosphate (PIP) cycle**, which is set off by the binding of the sperm to the egg cell membrane. Once the PIP cycle is activated, there is a sudden spike in calcium levels within the egg due to the release of calcium from the smooth endoplasmic reticulum. This spike in calcium levels causes hundreds of vesicles (the **cortical granules**) housed in the cortical cytoplasm of the egg to fuse with the egg cell membrane and to empty their contents into the perivitelline space between the egg cell membrane and the vitelline envelope. The released contents of the cortical granules swell, lifting the vitelline envelope away from the egg cell surface, and tan the envelope, making it tough and impenetrable by other sperm. The PIP cycle also causes a rise in internal pH by activating a sodium–hydrogen ion pump. Sodium ions are pumped into the egg while hydrogen ions are pumped out. The resulting rise in internal pH activates the egg, causing protein synthesis to start and the egg to begin its development.

The details above are sophisticated but not beyond your manipulation with relatively simple reagents. Think hard about what you could do to interfere with one or more aspects of the events of fertilization.

Instructions for normal fertilization

The same instructions can be used for sea urchins or sand dollars, except where noted. You will not be able to sex the animals until they have begun to spawn. Invert five or six animals over dry watch glasses or petri dishes. The five gonadal openings are on the aboral side, and as the animal spawns, gametes will be shed into the dish. On the oral side, you will see the hard white mouthparts ("Aristotle's lantern") surrounded by a tough leathery peristomial membrane. You will be injecting through this membrane into the perivisceral cavity. The length of the needle you choose should be long enough to penetrate into the cavity but not much longer. (When spawning sand dollars, a much shorter needle is used than when spawning sea urchins.) Use a syringe to inject 1–2 ml (0.5–1 ml for sand dollars) of isotonic KCl ($0.53 M = 3.9\%$). This will cause the smooth muscles of the gonads to contract and spawn their gametes. After 2–5 minutes, repeat the injection. This gives a heavier spawning.

As soon as spawning begins (within minutes of injection if the animals are ripe), check the color of the gametes. This will tell you the sex of the animal. Sperm are creamy white; eggs are yellow, pink, or dark red (depending on the species).

For males Immediately pour off the first sperm to get rid of perivisceral fluid, since this will interfere with the sperm's ability to fertilize. Then allow the animal to shed into the watch glass or petri dish without diluting the sperm. This is called **dry sperm**. Dry sperm will be good for 6–10 hours at room temperature and even longer if kept in the refrigerator.

For females Allow the female to shed eggs into seawater by placing her, inverted, on top of a beaker of seawater. The beaker should be small enough that the animal does not fall in. The beaker must be full enough so that seawater touches the aboral side of the animal. As the animal sheds, the eggs will drift down through the seawater and settle in the bottom of the beaker. After shedding is completed, decant off the seawater and add fresh. Do this twice. This washes the eggs of perivisceral fluid which interferes with fertilization. (*N.B.:* Sometimes artificially spawning sea urchins will cause parthenogenetic activation of the eggs. Examine egg batches for raised fertilization envelopes, which indicates parthenogenetic activation. These eggs could be used in the parthenogenesis section of the chapter.)

To fertilize the eggs, first make a **standard sperm suspension** of 1 drop of dry sperm in 10 ml of seawater. This suspension must be used within 20 minutes and then discarded. Dry sperm are relatively inactive. Diluted, however, they become very active and quickly use up their energy stores. Use sterile procedure and sterile glassware if you are keeping the eggs for culturing. Add two drops of standard sperm suspension to 10 ml of seawater containing eggs. Repeat after two minutes. After 10 minutes, decant off the seawater and add fresh. This culture can be kept for observing normal development. Know the species you are using, and decide on an appropriate temperature (or range of temperatures) for rearing (see Table 6.1). For long-term cultures, refrigerator temperatures are usually adequate. The seawater for culturing should be sterilized by being filtered through a 0.22-μm porosity filter. Streptomycin can be added (2 mg/liter) to retard bacterial growth, but is normally not needed in cultures maintained at cold temperatures.

Watching Fertilization

You can watch fertilization under the microscope by placing eggs on a slide and introducing sperm from one side. One way to do this is to place a large drop of egg suspension on a microscope slide. Put the slide on the microscope stage and have a footed coverslip close at hand. (A **footed coverslip** is made by nicking the corners of the coverslip against some hard paraffin, so that crumbs of paraffin remain attached at each corner. This will give just enough spacer between the slide and the coverslip to avoid crushing the eggs.) Then put a drop of sperm suspension on the slide close to but not touching the drop of eggs. Place the footed coverslip on the slide so that it covers both drops. This will cause the two drops to mix. Immediately focus on eggs that are mature. (Remember, those with large germinal vesicles are immature.) Use a clock with a second hand to time the events that you see. You should be able to determine the exact spot of sperm entry, which will be marked by the fertilization cone (Figure 6.1). Time the raising of the fertilization envelope. Do you see any eggs in which the fertilization envelope starts to rise but doesn't finish? Record any variations you see. Once the fertilization envelope is raised, do sperm continue to be attracted to the egg? Do the sperm bounce off the fertilization envelope or stick to it? Enter your answers in your notebook, along with any other observations you make.

If you can find some immature eggs, place these separately on a slide, and watch as you introduce sperm from the side. Are the sperm attracted to the egg? Compare this with sperm behavior near a mature egg.

Prepare a slide of a fertilized egg, place a footed coverslip over it, and look at it under a 40× objective. Close down the iris diaphragm on your microscope to increase contrast. Focus on sperm that are caught in the jelly surrounding the egg. Can you see any acrosomal processes? Why would you expect to see them? Enter your answer in your notebook, and diagram what you see. Do not go to oil immersion, since this will not work using these wet mounts and will only mess up the microscope by getting seawater on the objective lens.

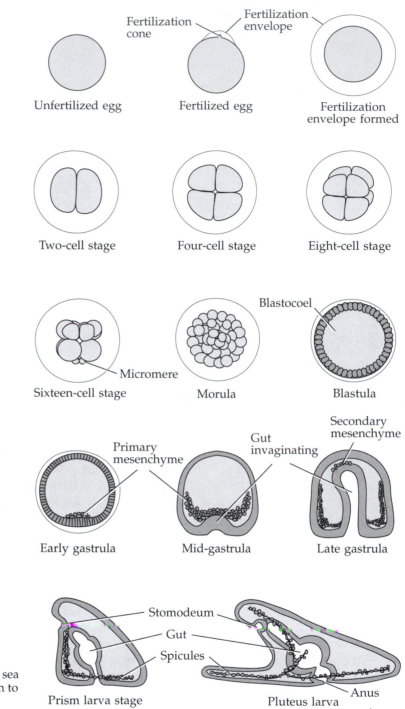

Figure 6.1
Developmental sequence of the sea urchin embryo from fertilization to the pluteus larval stage.

Put a drop of sperm suspension on a slide, and place a coverslip over it. Observe this under a 40× objective, and close down the iris diaphragm to increase contrast. Diagram the sperm. Do you see any acrosomal processes on these? Why? Record your answers.

Interfering with Fertilization

Many of the events important to fertilization depend upon the jelly surrounding the egg. What do you think would happen if you removed this jelly? You can remove it by exposing the eggs to seawater that has been adjusted to a pH of 5.0 by the addition of 0.1N HCl (added dropwise while measuring the pH with a pH meter). Swirl the eggs and return them to normal seawater after 2 minutes. When sperm are added to these eggs, how do the sperm behave? Does fertilization occur? What did you expect would happen? Enter your questions, answers, and observations in your notebook.

Remember that certain events in fertilization rely on sodium pumps in which sodium is being pumped into the egg. If sodium were unavailable to the egg, what might the results be? Design several questions you could ask, and try the fertilization procedures using sodium-free seawater. What do you think your results will be? What are your results? How might you explain any discrepancies between the expected and the actual results? Enter your questions and answers in your notebook.

Sodium-free seawater

Glycerol	84.60 ml	*Make up to 1 liter with distilled water.*
$MgCl_2 6 \cdot H_2O$	11.20 gm	
KCl	0.77 gm	
$NaHCO_3$	0.21 gm	
$CaCl_2$	0.12 gm	

Remember that calcium also is extremely important in fertilization. The calcium ionophore A23187 transports calcium ions across a cell membrane, causing a sudden increase in free calcium ion levels within the cell. What do you think the effect would be of adding A23187 to unfertilized eggs? Explain your reasoning. (For experiments, 1 mg A23187 is dissolved in 1 ml of dimethyl sulfoxide—DMSO—to make a stock solution. This can be stored in the freezer. The stock solution is diluted 1:100 in seawater for use. All operations should be done in a hood, wearing gloves, and using bench paper that is discarded after use. DMSO passes freely through the skin and will carry along with it anything that it is attached to.)

If you are following these cultures beyond class period, do not leave them in sodium-free seawater or DMSO solution. Return the embryos to normal seawater. You can do this by swirling the dish to concentrate the embryos. Then pipette them into fresh seawater. Do this several times to wash them clean of their previous solutions. Store them at a temperature appropriate for the species being used.

Parthenogenesis

Parthenogenesis, the development of an egg without fertilization, can be induced in the sea urchin in a number of ways. Remember, all that has to be achieved to activate the egg is an assault that will cause a rise in pH. (This is caused by activating the PIP cycle in normal fertilization.) In the past, numerous methods have been tried, often successfully. Hypertonic salt solutions as well as hypotonic solutions have worked, as have a pricking of the cell membrane and even violent shaking of the eggs. No one knows yet exactly how each method works, but it would be reasonable to suppose that each causes an activation of the PIP cycle.

We will use aged eggs (4–6 hours) since these often work better than freshly spawned eggs. Why do you think this is the case? Enter your answer in your notebook. Try putting several milliliters of an aged-egg solution in a glass vial and shaking it in your hand. Also try exposing the eggs for 10 minutes to a hypertonic solution of seawater: 2.2 ml of 2.5 M (14.6%) NaCl in 7.8 ml of normal seawater. The hypertonic method was developed by Ernest Everett Just in the early 1900s. (E. E. Just [1883–1941] was an eminent embryologist, a pioneer in the study of fertilization, and an African-American who suffered discrimination in this country. Unable to find a supportive atmosphere at home, he fled to Europe to pursue his work. Forced to return by the outbreak of World War II, he came home a sick man, and died, too young, of a painful intestinal neuralgia and pancreatic cancer. He's the only embryologist ever to be honored on a U.S. postage stamp [issued Feb., 1996].)

View the treated eggs under a dissecting scope. What are you looking for that will indicate activation of the eggs? Many times, you will see the elevation of the fertilization envelope. What does this tell you? Under conditions of normal fertilization, cleavage occurs within the first hour at room temperature. Leave some of your eggs at room temperature and place some in the refrigerator. These can be checked at intervals. If cleavage does occur (congratulations!), your eggs will be among the lucky 1% that can be expected to be parthenogenetically activated. Even those that are activated, however, rarely make it past the blastula stage. Why? Suggest an answer and enter it in your notebook.

If you plan on keeping these cultures beyond the class period, again, it is important that you remove them to normal seawater. Store them at a temperature appropriate for the species being used.

Cleavage, Gastrulation, and Larval Stages

The first several cleavages can be observed during the laboratory period. But you will have to use your sterile cultures to observe the later stages of development.

The stages and pattern of cleavage can be seen in Figure 6.1. Cleavage is **holoblastic** (the entire egg cleaves) and **radial** (the cleavage planes are parallel or at right angles to the animal-vegetal pole). The first cleavage takes about an hour at room temperature, and subsequent cleavages occur about every half hour. At the 16-cell stage, a small group of **micromeres** are cleaved at the vegetal pole. These are the **primary mesenchyme** cells and will be the first to show gastrulation movements. At 5–6 hours, the embryo is at the **blastula stage**, and by 7–8 hours, the embryo has hatched out of its fertilization envelope and is spinning around the dish using its cilia for locomotion. This is called a **hatched blastula**.

Gastrulation begins at about the hatched blastula stage. Primary mesenchyme cells first migrate into the blastocoel and then form a necklace of cells that will secrete the skeletal supports for the larva, the **spicules**. The spicules are first tripartite rods, and they eventually branch to have several arms. Gastrulation continues by **invagination** of the vegetal plate to form the **archenteron** (meaning "ancient gut"). The forming gut will elongate, capped by a loose collection of cells, the **sec-**

ondary mesenchyme. The secondary mesenchyme aid in elongating the gut with their contractile filopodia, pulling the archenteron toward the far wall of the blastocoel and guiding it to its final destination. They later disperse to form mesodermal organs. As the gut is developing, the embryo goes through a **prism larval stage** (between 18–20 hours), looking like an exquisite rotating jewel, and finally becomes a **pluteus (echinopluteus) larva** (at about 22–24 hours). The pluteus is an ornate organism, projecting long, delicate arms supported by branched spicules.

The timing of development will vary considerably with temperature. You can keep cultures in the refrigerator, at room temperature, or at any other temperatures the laboratory can supply. As soon as you have swimming blastulae, I suggest catching these in a sterile pipette and transferring them to a fresh culture dish with sterile seawater. This will avoid bacterial contamination from the decomposing embryos that didn't make it. Make diagrams of any stages you are able to observe. If you are lucky enough to get larvae, be sure to look at them under the microscope using polarized light. As the larvae rotate, their spicules briefly align with the polarizers with each turn, flashing beautifully for you.

Compare the rates of development at the different temperatures you used by presenting your data in chart and graph form. Notice also (from the cultures that don't make it) that you are establishing viable temperature ranges for the species you are using. You can see from Table 6.2 that complete timetables of sea urchin or sand dollar development are hard to come by. If you develop a complete timetable for one or more temperatures, this is publishable work.

Table 6.2 Developmental timetables for sea urchin and sand dollar embryos at various temperatures

Stage	Strongylocentrotus droebachiensis 4°C	Strongylocentrotus droebachiensis 8°C	Strongylocentrotus purpuratus 10°C	Echinarachnius parma 15°C	Arbacia punctulata 20°C	Arbacia punctulata 23°C
Fertilization	0	0	0	0	0	0
2-Cell stage	5 hrs	3 hrs	3.5 hrs	1.5 hrs	1 hrs	50 min
4-Cell stage	8 hrs	5 hrs	5 hrs	3 hrs	1.8 hrs	78 min
8-Cell stage	10.5 hrs	6.5 hrs	6 hrs	4 hrs	2.4 hrs	1.71 hrs
16-Cell stage	14 hrs	8.5 hrs	8.5 hrs	5.5 hrs		2.25 hrs
32-Cell stage	18 hrs	11 hrs				2.78 hrs
Morula				6.5 hrs		4 hrs
Early blastula				7 hrs	6 hrs	4.5 hrs
Mid-blastula				8 hrs		6 hrs
Hatching blastula	50 hrs	30 hrs	27–28 hrs	12 hrs	10 hrs	7–8 hrs
Early gastrula				25 hrs		12–15 hrs
Mid-gastrula			50 hrs			
Late gastrula			57 hrs	31 hrs		17 hrs
Prism stage	6.5 days	4 days	80 hrs	48 hrs		18 hrs
Early pluteus	11 days	7 days	4.7 days	50 hrs	48 hrs	20 hrs
Late pluteus				72 hrs		24 hrs
Metamorphosis		7–22 wks	63–86 days			5–16 wks

Sources: S. droebachiensis and S. purpuratus after Strathmann, 1987; *E. parma* after Karnofsky and Simmel, 1963; *A. punctulata* at 20°C after Costello et al., 1957; *A. punctulata* at 23°C after Harvey, 1956.

Accompanying Materials

Tyler, M. S. and R. N. Kozlowski. 2000. *Vade Mecum: An Interactive Guide to Developmental Biology*. Sinauer Associates, Sunderland, MA. "Sea Urchin." This chapter of the CD provides video of the living embryo, with details on the raising of the fertilization envelope, color-coding of germ layers, and an artistic rendition of metamorphosis based on Ethel B. Harvey's original photographs.

Gilbert, S. F. 2000. *Developmental Biology*, 6th Ed. Sinauer Associates, Sunderland, MA. Chapters 7 and 8. You will find in these chapters excellent discussions of fertilization, the PIP cycle, and development of the sea urchin to the pluteus larval stage.

Fink, R. (ed.). 1991. *A Dozen Eggs: Time-Lapse Microscopy of Normal Development*. Sinauer Associates, Sunderland, MA. Sequence 1. This shows sea urchin development from fertilization through spicule formation.

Selected Bibliography

Carroll, D. J., D. Albay, M. Terasaki, L. A. Jaffe and K. R. Foltz. 1999. Identification of PLC γ-dependent and -independent events during fertilization of sea urchin eggs. *Dev. Biol.* 206: 232–247. This paper sorts out which events of fertilization are caused by the rise in calcium within the egg due to activation of the PIP cycle. These include the cortical granule reaction, the intracellular rise in pH, MAP kinase dephosphorylation, DNA synthesis, and the onset of cleavage.

Costello, D. P., M. E. Davidson, A. Eggers, M. H. Fox and C. Henley. 1957. *Methods for Obtaining and Handling Marine Eggs and Embryos*. Marine Biological Laboratory, Woods Hole, MA. This superb book, now out of print, is worth its weight in platinum to any marine invertebrate embryologist. For each species covered, descriptions are given of where and when to collect the animals, how to obtain gametes, and how to culture the embryos.

Czihak, G. (ed.). 1975. *The Sea Urchin Embryo: Biochemistry and Morphogenesis*. Springer-Verlag, Berlin. This is now a classic, a rich collection of papers on sea urchin development.

Ernst, S. G. 1997. A century of sea urchin development. *Amer. Zool.* 37: 250–259. A beautiful summary of the early research on sea urchins, including a short analysis of some of E. E. Just's contributions.

Epel, D. 1977. The program of fertilization. *Sci. Amer.* 237(5): 128–138. This article, though now somewhat dated, is a beautifully illustrated, clearly written explanation of fertilization in sea urchins. It is well worth getting from Scientific American Offprints.

Foltz, K. R., 1995. Gamete recognition and egg activation in sea urchins. *Amer. Zool.* 35: 381–390. This is an excellent summary of fertilization in sea urchins.

Harvey, E. B. 1956. *The American Arbacia and Other Sea Urchins*. Princeton University Press, Princeton, NJ. This tome, written by Ethel Harvey at a time when few women were allowed to succeed in science, is an extremely precise, exhaustive compilation of information on *Arbacia*. Her photographs of development are among the most complete to be had.

Karnofsky, D. A. and E. B. Simmel. 1963. Effects of growth inhibiting chemicals on the sand dollar embryos Echinarachnius parma. *Prog. Exp. Tumor Res.* 3: 254–295. An excellent description of sand dollar development, this paper, found unexpectedly in a journal on tumor research, illustrates how grant money for medical research still can be utilized for studying normal development.

Manning, K. 1983. *Black Apollo of Science: The Life of Ernest Everett Just*. Oxford University Press, New York. A compassionate biography of an outstanding embryologist. It is E. E. Just who wrote, "We feel the beauty of Nature because we are part of Nature and because we know that however much in our separate domains we abstract from the unity of Nature, this unity remains." This biography of Just reminds us that scientists should always see the beauty of the whole that binds together the parts that we dissect.

Pearse, J. S. and R. A. Cameron. 1991. Echinodermata: Echinoidea. In A. C. Giese, J. S. Pearse and V. B. Pearse (eds.), *Reproduction of Marine Invertebrates. Volume VI: Echinoderms and Lophophorates*. Boxwood Press, Pacific Grove, CA, pp. 513–662. This series on invertebrate reproduction is an astonishing collection of information. The chapter on echinoids gives tables of spawning times for a long list of species and provides an excellent summary echinoid development.

Sherwood, D. R. and D. R. McClay. 1999. LvNotch signaling mediates secondary mesenchyme specification in the sea urchin embryo. *Development* 126: 1703–1713. The work of David McClay and his collaborators is an excellent example of how to combine old-fashioned observation with modern molecular techniques. Their work has made significant advances in our understanding of gastrulation. This paper summarizes a portion of that work.

Strathmann, M. F. 1987. *Reproduction and Development of Marine Invertebrates of the Northern Pacific Coast: Data and Methods for the Study of Eggs, Embryos, and Larvae*. University of Washington Press, Seattle. This extremely useful volume is the modern version for the Pacific coast of the classic volume for the Atlantic coast by Costello et al. (cited above).

Stricker, S. A. 1999. Comparative biology of calcium signaling during fertilization and egg activation in animals. *Dev. Bio.* 211: 157–176. This gives a concise review of what is known about the role of calcium in fertilization in all groups of metazoans studied so far.

Wessel, G. M. and A. Wikramanayake. 1999. How to grow a gut: ontogeny of the endoderm in the sea urchin embryo. *BioEssays* 21: 459–471. A well-written summary of gastrulation movements in the sea urchin and the mechanisms of these movements, this review also covers a great deal of the history of this field.

Wilt, F. H. and N. K. Wessells (eds.). 1967. *Methods in Developmental Biology*. Thomas Y. Crowell, New York. This is one of my favorite techniques books in developmental biology. For every species covered, it has a wealth of information in a compact space. If you ever see it on a "for sale" list, don't let it slip away.

Wray, G. A. and D. R. McClay. 1989. Molecular heterochronies and heterotopies in early echinoid development. *Evolution* 43: 803–813. The study of evolution and embryology should be inexorably linked but seldom are. This concise paper shows how the study of development can advance our understanding of evolution.

Vacquier V. D. and G. W. Moy. 1997. The fucose sulfate polymer of egg jelly binds to sperm REJ and is the inducer of the sea urchin sperm acrosome reaction. *Dev. Biol.* 192: 125–135. This paper discusses methods for dejellying sea urchin eggs in addition to reconfirming that a fucose-containing polysaccharide activates the acrosome reaction in the sea urchin sperm.

Suppliers

Connecticut Valley Biological Supply Co., Inc.
P.O. Box 326
82 Valley Road
Southampton, MA 01073
1-800-628-7748

Stongylocentrotus droebachiensis
Instant ocean

Gulf Specimen Marine Laboratories, Inc.
P.O. Box 237
Panacea, FL 32346
1-850-984-5297
www.gulfspecimen.org

Lytechinus variegatus
Arbacia punctulata (June-August)

Marine Biological Laboratory
Woods Hole, MA 02543
1-508-548-3705, ext. 325
www.mbl.edu

Stongylocentrotus droebachiensis (November–March)
Arbacia punctulata (mid June–mid August)
Echinarachnius parma

Pacific Bio-Marine Laboratories, Inc.
P.O. Box 536
Venice, CA 90291
1-310-677-1056

Sea Life Supply
740 Tioga Ave.
Sand City, CA 93933-3034
1-831-394-0828
www.sealifesupply.com

Stongylocentrotus purpuratus (late November–June)
Stongylocentrotus fransiscanus (late November– June)
Lytichinus pictus (May–September)

Any good biological supply company, such as:
Fisher Scientific
585 Alpha Dr.
Pittsburgh, PA 15238
1-800-766-7000
www.fishersci.com

Salts for artificial and Ca/Mg-free seawater
Sterilizing filters (0.22-µm pore size)
Sterile filter systems (disposable and reusable)

Any good chemical supply company, such as:

Sigma Chemical Company
P.O. Box 14508
St. Louis, MO 63178
1-800-325-3010
www.sigma-aldrich.com

DMSO
Calcium ionophore A23187
Streptomycin

7

Sea Urchin Development
Effects of Ultraviolet Radiation[1]

Ultraviolet Radiation in Earth's Atmosphere

What could be more enticing than a sun-drenched beach? Yet that inviting sunlight harbors great potential harm for us, as well as any organism exposed to the damaging ultraviolet rays that pass through our atmosphere. It has been known for some time that ultraviolet light is damaging, and indeed you may have already used it in the laboratory as a germicidal agent. Ultraviolet light kills—and it does so primarily by damaging DNA. So how does any organism survive exposure to ultraviolet light? And how much of a threat are the increasing levels of ultraviolet radiation that are reaching Earth's surface because of the damage we have caused to the protective atmospheric ozone layer?

The damage caused by **ultraviolet** (UV) radiation depends on its wavelength (Figure 7.1). UV wavelengths fall between 200 and 400 nanometers (nm). Above this is visible light, also called **PAR** (photosynthetically active radiation; **400–700 nm**) that is not considered harmful. Within the UV range, the longest wavelengths, which constitute **UV-A (320–400 nm)**, are the least harmful to biological systems. They are also the type of UV radiation that passes most easily through Earth's atmosphere. Only about 3% of the sunlight reaching Earth's surface is in the ultraviolet range, and most of this is in the UV-A range. **UV-B (280–320 nm;** though some sources use 290–320 nm) is significantly more harmful, but most of the UV-B coming from the sun (especially wavelengths below 295 nm) is blocked by the ozone in Earth's upper atmosphere (the stratosphere). Only about 0.25% of the sun's light reaching Earth's surface is UV-B. **UV-C (200–280 nm)** is the most highly dangerous to organisms; however, because of its absorption by oxygen in the atmosphere, UV-C never reaches Earth's surface. It is UV-B radiation, therefore, that is of primary concern to biological organisms, and anything that would shift the absorption of UV-B to allow more of this wavelength to reach the Earth's surface should be of major concern to us.

This is precisely what has been happening due to a reduction in the protective ozone shield of the upper atmosphere. This shield has been reduced in recent years by the production of **chlorofluorocarbons (CFCs**; these include freon, chloroform,

---

[1]With thanks to Nikki Adams, Department of Biological Sciences, University of Maine, for her help in developing this chapter.

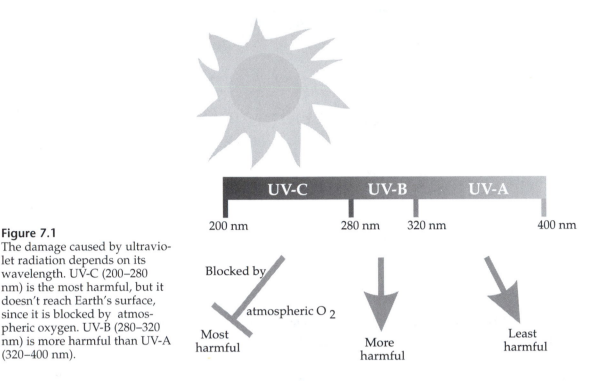

Figure 7.1
The damage caused by ultraviolet radiation depends on its wavelength. UV-C (200–280 nm) is the most harmful, but it doesn't reach Earth's surface, since it is blocked by atmospheric oxygen. UV-B (280–320 nm) is more harmful than UV-A (320–400 nm).

carbon tetrachloride, various cleaning agents of computer circuit boards, and propellants for aerosol cans). CFCs carry chlorine into the upper atmosphere where the chlorine destroys ozone. The increase in CFCs has created such severe ozone depletion that as much as 5% of the ozone shield has disappeared in less than a 10-year period. Even if we were to completely stop the production of CFCs today, it would still take 30 years before the CFCs that are presently in the lower atmosphere finally worked their way into the upper atmosphere. And it would take 70 years beyond that before the ozone levels were restored just to present-day levels. Major efforts toward the elimination of CFCs were made in 1987 when 24 countries signed the Montreal Protocol, and additional countries have signed since then. Even so, there are major **ozone "holes"** that have developed in the upper atmosphere, and these grow and shrink in size with seasonal regularity, becoming their largest in the winter months. Ozone depletion is a global problem, although it is most severe over Antarctica where one ozone "hole" at its largest is the size of the United States.

The effects of this ozone depletion are likely to be most devastating to organisms that are exposed to sunlight during their developmental stages. These include many freshwater and marine species, as well as terrestrial species such as amphibians that return to water for reproduction. Although it used to be thought that UV-B couldn't penetrate into water very far, it has since been learned that UV-B radiation can reach to depths of 20 meters, and UV-A radiation can penetrate even farther. So, many of the organisms that we used to think were protected from increased UV exposure instead are found to be threatened.

Biological Effects of UV Radiation

The damage caused by UV radiation that is environmentally relevant includes that caused by UV-A and UV-B. On an organismal level, this damage results in inhibition of photosynthesis, sunburn, photoaging of the skin, skin cancer, cataracts, damage to the neural retina, and depression of the immune system. On a cellular level, the damage can cause the alteration of DNA, proteins, and lipids, either through direct or indirect effects.

UV-A radiation causes damage indirectly by creating **reactive oxygen species**, such as superoxide anions (O_2^-) and hydrogen peroxide (H_2O_2). These highly reactive molecules cause damage by oxidizing proteins, lipids, and DNA. This oxidation can cause poorly functioning enzymes and structural elements of the cell, and it can result in mutations through alteration of the DNA, which can ultimately lead to cancer.

UV-A

UV-B radiation causes direct damage to proteins and DNA by creating the formation of **pyrimidine dimers** and **crosslinks** between DNA and protein. This form of radiation causes significant DNA damage, delayed cell divisions, and is responsible for most of the ill effects listed above as the effects of UV light on organisms.

UV-B

Researchers are beginning to investigate the extent of the effects of increased UV radiation caused by the thinning of our protective atmospheric ozone layer. Already, it has been shown that increased levels of UV-B radiation can be particularly damaging to gametes and embryos. Any organism whose gametes or embryos are exposed to sunlight are likely to be susceptible and at risk.

Mechanims That Protect Against Damage from UV

Organisms are not completely unprotected against the damage caused by UV light, and some have greater amounts of protection than others. Protective mechanisms include natural sunscreens, such as **blue pigments,** which reflect short-wavelength radiation, and pigments and other compounds that absorb UV-B, such as **mycosporine-like amino acids** (**MAAs**). MAAs are found in photosynthetic algae, fungi, and bacteria, and they can accumulate in non-photosynthetic organisms that feed on the MAA-rich organisms.

Physiological mechanisms also exist for repairing the DNA damage inflicted by UV light; these rely on DNA repair enzymes such as **photolyase**. These enzymes repair the UV-induced DNA dimers. Interestingly, these are **photorepair mechanisms**, being activated by light in the visible range and down into the UV-A range. Eggs and embryos can vary in the amounts of photolyase they contain, and these variations can be species-specific. In several amphibian species from the western United States, the species that showed the most deformities induced by UV-B radiation exposure were also the species with the lesser amounts of photolyase (Blaustein et al., 1994; Blaustein et al., 1998).

Using Sea Urchins as a Model Organism for UV Studies

Sea urchins can be used to study damage induced by ultraviolet radiation. They are exposed in nature to UV radiation during all stages of their life cycle, and any analysis of their sensitivity can be used as a measure of the extent of the threat that increased UV-B exposure poses. Although the adults would seem to be well protected from UV radiation by their heavily calcified tests, their gametes and embryos are transparent and poorly protected.

It has long been known that UV radiation is harmful to sea urchin development. As early as 1877 (Downes and Blunt), UV radiation was shown to induce mitotic delay in cells. These results were verified and and shown to apply to sea urchin embryos as well (e.g., Giese, 1946; Rustad, 1971). It has also been shown that UV radiation has adverse effects on sperm, causing sperm to agglutinate, be less motile, and have reduced fertilizing abilities. In most of these earlier studies, however, the emphasis was not on the threats of environmental exposures to UV radiation, and the wavelengths used were often in the UV-C range, a range that isn't environmentally relevant. More recently, however, the ill effects of UV light in environmentally relevant ranges have been documented (e.g., Adams and Shick, 1996; Anderson et al., 1993). These studies show that environmental UV light can cause cleavage delay and a number of developmental abnormalities, such as blastulae that are filled with an abnormally large number of cells, exogastrulae, and bizarrely shaped pluteus larvae with extra spicules. The embryos have proven to be most susceptible to damage when they are exposed during early stages (up to the formation of the blastula) rather than during later stages.

The embryos are not without some protection against UV radiation. Adams and Shick have documented that the accumulation of natural sunscreens (the MAAs discussed earlier) protect the eggs and embryos. These MAAs, found in the algae that is consumed by the adult animal, are stored in the eggs, providing a certain level of protection, depending upon the amounts accumulated. Therefore, by selecting algae high in MAAs, an adult animal can influence the protection from UV radiation that it provides the next generation.

Setting Up Your Experiments

The questions you can ask of sea urchin gametes and embryos about the effects of UV are limited only by your imagination.

- Is the sperm or the egg more susceptible to UV radiation?
- What stages of development are most susceptible to damage by UV radiation?
- Will exposure to differing lengths of daylight affect development?
- Is UV-B or UV-A radiation more harmful to these embryos?
- Is the combined effect of UV-B and UV-A radiation exposure greater than the effect of exposure to only one of these ranges?
- Do antioxidants, such as vitamin C, counteract the effects of UV-A radiation and thus reduce the formation of reactive oxygen species?
- Does exposure to just PAR (visible light) cause any defects?
- Does exposure to PAR mitigate any of the ill effects caused by UV radiation exposure?
- Do the screens for filtering out UV light that we rely on to protect us—such as sunglasses and sunscreen lotions for our skin—actually work?

In getting started, you will want to refer back to Chapter 6, "Echinoid Fertilization and Development," to review how to obtain gametes from the sea urchin, and how to store these gametes until you are ready to use them. Remember that the eggs should be washed several times in seawater to remove any perivisceral fluid and that the sperm should be stored as "dry sperm." Eggs should be maintained at temperatures appropriate for the species. Sperm can be stored in the refrigerator. You should be using these gametes as soon after spawning as possible. Remember that aged eggs (4 hours old or older) are more susceptible to parthenogenesis.

In preparing your cultures, it is best to use artificial or filtered seawater, if possible. This avoids the inclusion of dissolved organic material, which can become oxidized when exposed to UV light, forming reactive oxygen species such as hydrogen peroxide.

You can set up your cultures in petri dishes. You should put no more eggs in a dish than will form a monolayer, since you want to achieve equal exposure to the light conditions. To keep sperm from using up their energy with active swimming, they can be maintained as a thin layer of dry sperm on ice or as diluted sperm in cold (4°C) seawater. Cultures will have to be kept covered so that they don't lose water to the environment. The cover you choose will depend on the type of light you want to let through.

If you want to test the effects of UV-A and UV-B that are occurring naturally, you can set your cultures up outside to use natural sunlight. Alternatively, you can use a light that mimics the range of light in nature. To test the effects of UV-A and UV-B separately, you can use filters that block certain wavelengths of light, or you can use bulbs that emit only in certain wavelengths. These bulbs are discussed below.

In setting up your exposures, you can use a short exposure period (such as 30 minutes) followed either by dark conditions (to prevent photorepair mechanisms from working) or by visible light conditions that will allow for photorepair. Exposures also can be set to mimic natural light-dark cycles. Using sunlight will automatically give you a natural light-dark cycle. Alternatively, you can choose constant exposure, although you should be aware that some of these exposures could be exceedingly harmful to your embryos and may kill them.

Whatever your setup, remember that you should be keeping your gametes and embryos at a temperature that promotes development. You would have determined this temperature during your previous laboratory on sea urchin development. In many cases, this means that you have to keep your cultures cold. If using natural sunlight, you may need to keep your cultures sitting on ice that you replenish as needed. If using artificial light, again, maintain cultures on ice if needed, or preferably in a cold room or temperature-controlled water bath.

Special lightbulbs that can be used

Everyday room lights   Room lighting provides light in the PAR (visible) range. An incandescent bulb emits negligible levels of UV light. Fluorescent bulbs, however, emit measurable levels of UV-B light as well as PAR, and this factor should be considered in your experiments. (You could even test to see if there were any differences between exposure to a fluorescent bulb versus an incandescent bulb.) An unfiltered halogen bulb emits a continuous spectrum that reaches all the way into the UV-C range.

Grow lights   A grow light, a light commonly used to enhance growth of house-grown plants, provides light enhanced in the UV-A range without emitting light in the UV-B range.

Sunlamps   A sunlamp, such as those sold for home use (e.g., Sylvania FS20 sunlamp), provides light enhanced in the UV-B range. Sunlamps from 20 years ago

(275 watts) emit high levels of both UV-A and UV-B radiation.

Black lights   A black light, a common accessory to youthful parties (e.g., NEC T10, 20 watts), provides UV-A light.

UV 340 lamps   A UV 340 lightbulb supplies light in both the UV-A and UV-B ranges and mimics the light found in nature. These bulbs are commonly used to test paints for weathering in sunlight.

Germicidal lamps   A germicidal UV bulb emits light enhanced in the UV-C range. It is therefore not environmentally relevant, but it can be used to show the detrimental effects of this exceedingly harmful radiation.

Filters that can be used to exclude UV-A or UV-B

Whether you are using natural sunlight or a UV 340 bulb, you can test the differing effects of UV-A and UV-B light by filtering out one or the other using various plastics and glass that block specific wavelengths of light. These sheets can be placed over your culture dishes. Alternatively, if you are working out in the field with organisms in their natural environment, you can use containers or bags made from the material. You can place the embryos in the containers or bags and suspend them in the water where they are normally found.

Filters that block UV-B	Allows through UV-A and PAR	glass Xerox transparencies plastic milk bottle Mylar film
Filters that block UV-A and UV-B (UV opaque)	Allows through PAR	standard plexiglass "light-block" plastic milkbottle sunglasses (rated as 100% UV blocking) sunscreens (variable) UV opaque plexiglass
UV transparent filters	Allows through UV-A, UV-B, and PAR	plastic petri dish plastic sandwich bag plastic food wrap UV transparent plexiglass cellulose acetate

Filters that block UV-B

These will test the effects of exposure to UV-A and PAR.

Glass   Plate glass blocks almost all UV-B light while allowing through most UV-A light and PAR. For example, 1/8-inch plate glass blocks all UV-B while allowing through 76% of the UV-A light and PAR. A glass slide blocks all of the UV-B while allowing through 85% of the UV-A light and PAR. A finger bowl blocks all of the UV-B while allowing through 61% of the UV-A and PAR. (N.B.: Pyrex glass petri dishes do not block UV-B effectively.)

Xerox transparencies   The transparency films that are used for copier machines block 95% of UV-B while allowing through 77% of UV-A and PAR.

Plastic milk bottle   An ordinary plastic milk bottle blocks all UV-B light while allowing through 55% of the UV-A and 80% of PAR.

Mylar   A special type of Mylar film (Mylar type D fluoropolymer film) blocks all UV-B radiation.

Filters that block both UV-A and UV-B (UV opaque)

These will test the effects of exposure only to PAR (visible light).

Standard plexiglass   Standard plexiglass blocks all UV-B and most UV-A radiation while allowing PAR through.

"Light-block" plastic milk bottle   The special "light-block" plastic bottles designed to protect the vitamins in milk from breaking down block all UV-B light and virtually all UV-A, while allowing through only 14% of PAR.

Sunglasses   Your own sunglasses, if bought recently, come with a rating as to how much UV-A and UV-B they block. It would be interesting to test these as a filter, both for their protective effects on the embryos as well as for the reliability of their advertised claims.

Sunscreens   The sunsreens you buy to shield your skin may or may not be effective in blocking UV radiation. You could test them by applying a thin film on a UV-transparent substance such as a plastic petri dish or a plastic sandwich bag.

UV opaque plexiglass   A specially formulated plexiglass can be obtained (Plexiglas® G UF-3 acrylic) that blocks all of UV-A and UV-B but allows through PAR.

Filters that do not block either UV-A or UV-B (UV transparent)

These can be used as a control cover, so that all cultures are covered by a plastic sheet.

Plastic petri dish   A plastic petri dish allows through most UV-A (81%) and substantial amounts of UV-B (50%) radiation as well as PAR.

Plastic sandwich bag   The plastic bags you use for food storage allow through almost all UV-A, UV-B, and PAR.

Plastic food wrap   The plastic food wrap that you commonly wrap around your sandwiches allows through virtually all UV-A, UV-B, and PAR.

UV transparent plexiglass   A specially formulated plexiglass can be obtained (Plexiglas® G UVT acrylic) that doesn't block PAR, UV-A, or UV-B. It does block UV-C.

Cellulose acetate   Cellulose acetate, like UV transparent plexiglass, doesn't block PAR, UV-A, or UV-B, but it does block UV-C. Since it is less expensive than UV transparent plexiglass, it may be a preferred material to use.

Safety first

Remember that any bulb emitting UV light will be damaging to you. UV light is very damaging to your eyes: it can cause cataracts by damaging your lens, and blindness by damaging your retina. You must wear protective glasses, therefore, when exposed to UV light. If possible, avoid having the UV lights on when you are present. UV light also causes **photoaging** of the skin (making your skin leathery and lined), as well as **skin cancer**. So you should wear protective clothing as well whenever you risk exposure to UV light.

Monitoring Your Cultures

In monitoring your cultures, you can look for a number of things: fertilization delays, cleavage delays, and developmental delays, as well as abnormalities in any of the stages.

Record any differences that you see in your cultures as a percent of normal. Count 100 embryos, then record the number of abnormal/normal.

Data simply on the timing of development are valuable. Realize that an alteration in the rate of development can cause embryos to emerge as juveniles when natural conditions may not be optimal for them. You can record when your cultures reach the blastula stage; the appearance of a hatched blastula, early gastrulation, and late gastrulation; the stage of spicule formation; and so on. Always compare your experimental results with a control culture.

Developmental abnormalities are also very likely to occur, especially on exposure to increased levels of UV-B radiation. Look for any developmental abnormalities, such as a blastula stage that fills up with cells, exogastrulae, abnormally long cilia, and extra spicules.

The abnormalities you observe may deal with size differences. You can determine changes in size by measuring the diameters of cleavage-stage embryos and both the outer and inner diameters of blastula stages and later. This will allow you to determine if a smaller blastocoel is caused by a smaller embryo or by taller cells.

Keeping Up with the Issues

When it comes to dealing with the issue of UV radiation, it is important to avoid being part of the problem and to actively be part of the solution. Increased UV radiation levels affect all of us, and already we are experiencing the effects through increased levels of skin cancer and cataracts. One way to participate is to keep informed. Several websites are devoted to ultraviolet radiation monitoring and can keep you up to date on increased levels in your area as well as globally. For the United States, these include the National Ultraviolet Monitoring Center (http://oz.physast.uga.edu/) and the EPA's Ultraviolet Radiation Monitoring Network (www.epa.gov/uvnet). Another EPA site deals specifically with the depletion of the ozone layer (www.epa.gov/docs/ozone). It details how you can avoid adding to the problem by explaining what substances cause ozone depletion and what substances can be used as substitutes. It also lists regulations that are in place. A site that gives information on international efforts to protect the ozone layer, run by the United Nations, can be found at www.undp.org/seed/eanda/montreal.htm.

Accompanying Materials

Tyler, M. S. and R. N. Kozlowski. 2000. *Vade Mecum: An Interactive Guide to Developmental Biology.* "Sea urchin–UV." This chapter of the CD has interactive units on normal sea urchin development and on the developmental abnormalities in sea urchin embryos caused by UV-B exposure.

Gilbert, S. F., 2000. *Developmental Biology*, 6th Ed. Sinauer Associates, Sunderland, MA. Chapter 3. This chapter discusses the harmful effects of ultraviolet radiation on embryos.

Fink, R. (ed.). 1991. *A Dozen Eggs: Time-Lapse Microscopy of Normal Development.* Sinauer Associates, Sunderland, MA. Sequence 1. Shows sea urchin development from fertilization through spicule formation.

Selected Bibliography

Abney, J. R. and B. A. Scalettar. 1998. Saving your students' skin. Undergraduate experiments that probe UV protection by sunscreens and sunglasses. *J. Chem. Ed.* 75: 757–760. This article analyzes the UV-blocking capabilities of a number of commercial sunscreens and sunglasses.

Adams, N. L. 1998. Ultraviolet radiation affects development but not reproduction of green sea urchins. *Amer. Zool.* 38 (5): 160A. A short report examining the extent that UV radiation interferes with reproduction in the sea urchin.

Adams, N. L. 1999. The green sea urchin, *Strongylocentrotus droebachiensis*, covers itself in response to ultraviolet radiation. *Amer. Zool.* 39: 113A. A beautiful study showing that sea urchins protect themselves from UV radiation in part by covering themselves with shells and seaweed, or whatever is available to them.

Adams, N. L. and J. M. Shick. 1996. Mycosporine-like amino acids provide protection against ultraviolet radiation in eggs of the green sea urchin, *Stongylocentrotus droebachiensis*. *Photochem and Photobiol.*, 64: 149–158. A detailed analysis of MAAs in sea urchin eggs, this study reports that MAAs do protect these eggs from UV radiation.

Anderson, S., J. Hoffman, G. Wild, I. Bosch and D. Kabentz. 1993. Cytogenetic, cellular, and developmental responses in Antarctic sea urchins (*Sterechinus neumayeri*) following laboratory ultraviolet-B and ambient solar radiation exposures. *Antarctic J.* 28: 115–116. An analysis of the dangers that ozone holes over Antarctica pose for sea urchins.

Blaustein, A. R., P. D. Hoffman, D. G. Hokit, J. M. Kiesecker, S. C. Wall, and J. B. Hays. 1994. UV repair and resistance to solar UV-B in amphibian eggs: a link to population declines? *Proc. Natl. Acad. Sci, USA* 91: 1791–1795. This paper discusses the harm that UV-B radiation causes in amphibian eggs and the protective effects of photolyase in these eggs.

Blaustein, A. R., J. M. Kiesecker, D. P. Chivers, D. G. Hokit, A. Marco, L. K. Belden and A. Hatch. 1998. Effects of ultraviolet light on amphibians: Field experiments. *Amer. Zool.* 38: 799–812. An excellent review of the effects of UV radiation on numerous amphibian species.

De Gruijl, F. R. and P. D. Forbes. 1995. UV-induced skin cancer in a hairless mouse model. *BioEssays* 17 (7): 651–660. This is an excellent discussion of the hazards of UV radiation that humans are commonly exposed to.

Downes, A. and T. P. Blunt. 1877. Researches on the effect of light upon bacteria and other organisms. *Proc. Roy. Soc. Lond.* 26: 488–500. A lesson in history. As early as 1877, these researchers showed that UV radiation induced mitotic delay in cells.

Dunlap, W. C. and J. M. Shick. 1998. Ultraviolet radiation-absorbing mycosporine-like amino acids in coral reef organisms: a biochemical and environmental perspective. *J. Phycol.* 34: 418–430. A superb review of the protective value of MAAs in coral reef organisms.

Epel, D., K. Hemela, J. M. Shick and C. Patton. 1999. Development in the floating world: defenses of eggs and embryos against damage from UV radiation. *Amer. Zool.* 39: 271–278. A well-written article that discusses many of the questions surrounding UV radiation damage to embryos while focusing of the damage and protective devices in tunicate embryos.

Giese, A. C. 1946. Comparative sensitivity of sperm and eggs to ultraviolet radiations. *Biol. Bull* 91: 81–87. An early study that shows that both sperm and eggs are ill-affected by UV radiation. The radiation used was in the UV-C range.

Gulko, D., M. P. Lesser and M. Ondrusek. 1995. Introduction to materials and methods commonly used by participants in the 1994 H.I.M.B. Summer Program on UV Radiation and Coral Reefs. Ultraviolet Radiation and Coral Reefs, 1995. (D. Gulko and P. L. Jokiel, eds.) *HIMB Tech. Report* #41. UNIHI-Sea Grant-CR-95-03. Pp. 19–23. This very handy paper gives the nitty-gritty details about the sources and properties of materials used in UV radiation studies.

Kerr, J.B. and C. T. McElroy. 1993. Evidence for large upward trends of ultraviolet-B radiation linked to ozone depletion. *Science* 262: 1032–1034. An important paper verifying that depletion in the ozone of the stratosphere has caused increases in our exposure to UV-B radiation.

Mead, K. S. and D. Epel. 1995. Beaker versus breakers: how fertilization in the laboratory differs from fertilization in nature. *Zygote* 3: 95–99. This is a concise summary of the ecology of fertilization, including a discussion of the protective devices sea urchin eggs must have against ultraviolet radiation.

Milne, D. H. 1995. *Marine Life and the Sea*. Wadsworth Publishing Co., Belmont, CA. This textbook gives a concise description of ultraviolet radiation in our atmosphere and the destructive effects of CFCs on ozone.

Peak, M. J. and J. G. Peak. 1989. Solar-ultraviolet-induced damage to DNA. *Photodermatology* 6: 1–15. A sophisticated review of the damage that UV radiation causes to DNA.

Rustad, R. C. 1971. Radiation responses during the mitotic cycle of the sea urchin egg. In *Developmental Aspects of the Cell Cycle* (I. L. Cameron, G. M. Padilla and A. M. Zimmermann, eds.). Academic Press, New York, pp. 127–159. This is an excellent review of the specific damage that UV radiation has on eggs that results in cleavage delay. It also discusses the ill effects that UV radiation has on sperm.

Shiroya, T., D. E. McElroy and B. M. Sutherland. 1984. An action spectrum of photoreactivating enzyme from sea urchin eggs. *Photochem. and Photobiol.* 40: 749–751. An important reference showing that the range of light that activates photolyase repair mechanisms in sea urchin eggs is 313–500 nm with a maximum effect at 365 nm.

Suppliers

Cadillac Plastic
2855 Coolidge Highway, Suite 300
Troy, MI 48084
1-800-274-1000
www.cadillacplastic.com

Plexiglas® G UVT Acrylic, 6 mm thick (UVT—allows through UV-B, UV-A, and PAR)
Plexiglas® G UF-3 acrylic sheet, 6 mm thick (UVO— blocks all UV, allows through PAR only)

Allied Signal Plastics
P. O. Box 1205
Pottsville, PA 17901
1-800-934-5679
www.honeywell.com

Aclar 33c Fluoropolymer film, 127 microns thick 5 gauge (UVT—allows through UV-B, UV-A, and PAR)

DuPont Teijin Films
Barley Mill Plaza, Bldg 27
Lancaster Pike and Route 141
P. O. Box 80027
Wilmington, DE 19880-0027
www.dupontteijinfilms.com

AIN Plastics
249 E. Sandford Blvd.
P. O. Box 151
Mount Vernon, NY 10550
1-800-431-2451
www.tincna.com/ain1.htm

Mylar type D fluoropolymer film, 127 microns thick (5 mil) (Blocks UV-B—allows through UV-A and PAR)

Polycast Technology Corporation
70 Carlisle Place
Stamford, CT 06902
1-800-243-9002
www.polycast.com

Carolina Biological Supply Company
2700 York Road
Burlington, NC 27215
1-800-334-5551 or
1-910-584-0381
www.carolina.com

UV 340 bulbs (mimic the UV-A and UV-B range of nature)
UV-C (germicidal) lamps—not environmentally relevant

Science supply companies such as:
Ward's Natural Science Establishment, Inc.
P.O. Box 92912
5100 West Henrietta Road
Rochester, NY 14692-9012
1-800-962-2660
website: www.wardsci.com

UV safety goggles

chapter 8

Development of the Fruit Fly
Drosophila melanogaster

Life Cycle

The fruit fly *Drosophila melanogaster* is the familiar visitor on your overripe bananas and an organism of choice in genetics laboratories. As bridges between genetics and developmental biology are both built and traveled upon, it becomes imperative that developmental biologists study *Drosophila* to aid in the union of these two disciplines. Some of the major questions in developmental biology can only be answered with genetics. So we must learn about the geneticists' organisms and make them ours as well. An embryologist well-schooled in *Drosophila* development is both rare and valuable. Study this laboratory exercise well and it could make you some bucks in the future.

Drosophila is a **holometabolous** insect—that is, an insect that has a larval and a pupal stage prior to the adult stage. (Hemimetabolous insects, on the other hand, have nymph stages preceding the adult.) The adult *Drosophila* may live for more than 10 weeks. During this time, mating takes place. Fertilization is internal, and sperm are stored within the female's body in a **seminal receptacle** and the paired **spermathecae**. Females reach the peak of their egg production between the fourth and seventh day after their emergence. During this time, they lay eggs almost continuously at a rate of 50–70 eggs per day.

The eggs are approximately one-half mm in length, white, oval, and slightly flattened in lateral view (they look much like a kernel of rice). The ovum is surrounded by an inner, very thin **vitelline envelope** and an outer, tough extracellular coat called a **chorion**. At its anterior end two small filaments, extensions of the chorion, extend from the dorsal surface. These are **respiratory filaments** and serve for gas exchange, as their name implies. Eggs are laid half-buried in rotten fruit or the medium in your culture jars, and the filaments protrude into the air.

Eggs hatch in 22–24 hours at 25°C. The larva that emerges looks like a tiny worm and is called the **first instar larva**. It feeds on the substrate that the eggs were laid in and, after another 25 hours, molts into a larger wormlike form, the **second instar larva**. This feeds as well and, after about 24 hours, molts into the **third instar larva**. This is the largest of the larval forms. It feeds, but it also starts to climb upward out of its food, so that it will be in a relatively clean and dry area to undergo pupation. The third instar molts into a **pupa** after 30 hours. The pupa is stationary, and in its early stages is yellowish-white. As it develops, the pupa becomes progressively darker. During the pupal stage, the larva is **metamorphos-ing** into the adult fly, also called the **imago**. In doing so, it lyses most of the larval structures, although some larval organs are preserved. The larval nervous system, for example, is not lysed, but even it undergoes major restructuring; the Malpighian tubules (excretory structures), fat bodies, and gonads are kept as well. Most of the adult structures, however, form anew from two sets of cells that have been carried as undifferentiated, mitotic cells within the larva throughout its in-star stages: these are the **imaginal discs** (*imaginal* since they are for the *imago*) and the **histoblasts**.

Imaginal discs   These are small, almost teardrop-shaped packets of epithelial cells that will form the epidermal structures of the adult, such as the wings, legs, eyes, mouthparts, and genital ducts. Imaginal discs are carried around within the larva, growing in size but not differentiating. During the pupal stage, they evert and differentiate into their adult structures.

Histoblasts   These cells are found in small groups (nests) within the larva. They form the abdominal epidermis and the internal organs of the adult. They, too, grow by mitosis during the larval stages and then differentiate during the pupal stage. They are recognizable within the larva as clumps of small cells nestled among the huge differentiated polytene larval cells.

The pupal stage lasts for 3–4 days, after which the adult fly, or imago, emerges from the pupal case (**eclosion**). Adult male flies are sexually active within hours of emerging, females don't have ripe eggs until two days after eclosion, and the cycle begins again.

Pause for a minute. Think about what you've just read. It's weird! What does the *Drosophila* do during its life cycle? It has a larval form: a fully differentiated, feeding organism, that carries around, as extra baggage, cells that will replace it—cells that will become *another* fully differentiated, feeding organism. The larva is only a vessel, a nurturing culture environment for these cells that become the adult. At the appointed time, the larva self-destructs as these "passenger cells" differentiate. It is an astonishing way to make an adult.

Culturing Drosophila melanogaster

As you have undoubtedly noticed from the fruit basket that sat too long, *Drosophila* thrive on fermenting soft fruits. A very suitable culture medium, there-fore, is crushed banana. It provides all the necessary nutrients for both the larval and adult stages. The banana can be kept along with the flies in sterile pint jars with cotton or foam rubber plugs.

Another standard medium, commonly used by laboratories that raise *Drosophila,* is a cornmeal-molasses-agar mixture. While the batch brews, it fills the scientific hallways with the smells of Grandpa's favorite cookies.

Cornmeal-molasses-agar culture for **Drosophila**

water	420 ml	*Mix and boil water and agar 3–5 minutes.*
agar	4.5 gm	*Add unsulfured molasses and heat to*
unsulfured molasses	60 ml	*boiling again.*

cornmeal	49 gm	*Mix together cornmeal, brewer's yeast, and cold water in a separate container until all lumps are removed.*
brewer's yeast	6.5 gm	*Add cornmeal-yeast mixture to molasses-agar mixture. Boil 5 minutes, stirring constantly.*
cold water	145 ml	*Cool mixture to 60°C. Add propionic acid (as mold inhibitor).*
propionic acid	3.4 ml	*Pour culture medium 1-inch deep into sterile culture jars with sterile plugs. Pint milk bottles work well, but any widemouthed jar fitted with a plug made of cotton covered with cheesecloth or foam rubber should work well. Add a sprinkle of active baker's yeast (from a salt shaker) to each jar before adding flies.*

It is important when maintaining cultures not to overcrowd (about 100 flies per pint culture jar) and to subculture approximately every other day. This keeps the flies healthy, large, and mold-free.

Collecting Eggs

Collecting fertilized eggs is easy, but it is not easy to catch the very early stages of development, since eggs can be held within the female's uterus after fertilization for a period of time, even as late as just prior to larval hatching. When a female is laying rapidly, however, the uterus is being cleared fast, and eggs in their early stages of development can be obtained. To achieve this, it is best to use cultures of flies that are 5 days old. A female is within her peak laying period at this time and is laying eggs as quickly as one every 3 minutes.

Collecting chamber

A simple collecting chamber consists of an empty culture bottle (any wide-mouthed bottle will do) with wet toweling stuffed in the bottom (for humidity) and a cotton or foam rubber plug. Place approximately 40 pairs of flies in the chamber by inverting a culture bottle containing flies over the mouth of the empty bottle. Holding the bottles together, bang the empty bottom bottle against padding on a tabletop to cause the flies from the upper culture bottle to drop into it. Quickly replace the plugs of both bottles.

Use plastic spoons whose handles have been cut so they fit in the culture chamber without touching the plug. Put culture medium on the spoon (the same culture medium that the flies have been grown in), score it to make grooves (female flies like to lay their eggs in moist grooves), paint a light coating of baker's yeast suspension (a slurry of bread-making yeast in water) on the surface of the scored medium, and place one or two of these spoons in the collecting chamber with the flies. Since flies will outfly your quick fingers and escape from the bottle during this process, it helps to face the bottom end of the bottle toward a bright light. The flies will be attracted away from the mouth of the bottle. Place the bottle on its side so the medium won't slip off the spoons.

After about an hour, a suitable number of eggs will have been laid on the spoons, and they can be removed and replaced with fresh spoons containing medium. Again, attract the flies away from the bottle mouth with a bright light. If you prefer to have a number of different stages, including advanced stages, on a single spoon, leave the spoon in the collecting chamber for an extended period (up to 24 hours). Females produce the greatest number of eggs in the late afternoon and evening.

Mating behavior of adult flies

Before removing the spoons from the bottle, observe the adult flies through the glass, or place males and females in a small covered petri dish and observe them under the dissecting scope. To remove flies easily, first place the bottle in the refrigerator or keep it on ice for at least 20 minutes. This numbs the flies, and you can remove them without their escaping into the room. (Numbing with cold is an easier and safer way than anesthetizing with ether or CO_2.) You can distinguish males from females by looking for the black pigmentation on the posterior abdominal segment of the males; it is absent from females. Also, males have a shorter abdomen with six segments rather than eight, and they have a **sex comb** (a fringe of ten or so black, stout bristles) on the end of the first segment of the front legs.

Watch for **courtship behavior**. A female is very much in control of whether she is inseminated, being larger and stronger than a male. She must give an **acceptance signal** by slowing down, extruding her ovipositor, and spreading her wings, in order for mating to occur. There is no known incidence of rape among these organisms. A female rejects a male by kicking with her hind legs, fending with her middle legs, flicking her wings, producing a rejection buzzing sound by fluttering her wings, or moving away rapidly. If she has already mated, she also will extrude her genitalia to reject the male. A male courts anything that produces the right "taste" or "smell" (even other males if they are immature). He orients himself toward the female's head, taps her with his forelegs, "tasting" her to make sure she is the right species, and then pursues her when she moves, extending and vibrating one wing producing a **courtship song**. Though the "love song" of the *Drosophila* is species-specific, females do respond to the songs of other species as well. Females "hear" the song through their antennae; the **aristae** (feathery extensions of the antennae) augment the vibrations, and they are sensed by Johnston's organ in the second segment of the antenna. Later in the courtship, the male extends his proboscis to touch the female's genitalia. If all the active courtship of the male has stimulated the female enough to accept the male, the two mate with the male positioned on top of the female.

Sperm travel through the male penis into the female uterus and then swim into the female **seminal receptacle** and **spermathecae**, where they are stored for fertilization. During **oviposition**, the eggs emerge from the female's **ovipositor** posterior end first. The female oviposits preferentially on a moist food surface in a humid atmosphere. If the air is too dry, the female may feed, but she won't oviposit.

Observations of the Egg (Use Sterile Technique)

After a suitable waiting period, remove a spoon from the collecting chamber, and look at the surface of its medium under a dissecting scope. You should see a number of white eggs, often with just their respiratory filaments sticking out of the culture medium. You may find the eggs clustered in patches, laid preferentially on the moister regions of the medium. Scan the entire surface of the medium. Use a microknife or fine forceps to remove the eggs to a small petri dish containing insect Ringer's solution. (This is a balanced salt solution that will allow continued development of the egg until hatching.)

Insect Ringer's solution

NaCl	7.5 gm	*Make up to 1 liter with distilled water.*
KCl	0.35 gm	
CaCl$_2$	0.21 gm	

The chorion

Observe the eggs under a dissecting microscope and record your observations with diagrams. Your initial observations will have to be of superficial structures until you remove the chorion from the egg. Notice that the **chorion** is thick and tough, and that the two **respiratory filaments** and a **micropyle** are extensions of this chorion (Figure 8.1). The micropyle is a tiny channel that leads to the ovum, and it is through this channel that a sperm must swim in order to fertilize the egg. Though there have been reports of polyspermy in *Drosophila*, normally only one sperm is successful in making the trip. What do you think the purpose of the micropyle is? Record your answer in your notebook.

Now place several of the eggs in a drop of Ringer's solution on a slide and observe these under a compound microscope. Focus at the surface of the chorion as well as deeper. Place a small drop of toluidine blue over the eggs. This dye will help to show the ornamental markings of the chorion. These markings are beautiful. They are the impressions ("footprints") left by the **ovarian follicle cells** that deposited the chorion prior to ovulation.

The chorion is a complex structure, consisting of inner and outer laminae bounding a layer of tiny air pockets that connect anteriorly with an extensive

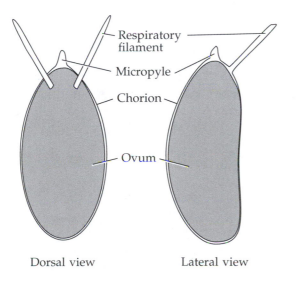

Respiratory filament

Micropyle

Chorion

Ovum

Dorsal view   Lateral view

Figure 8.1
Diagrams of the *Drosophila* egg.

meshwork of airspaces in the respiratory filaments. This design maximizes gas exchange and minimizes water loss. The respiratory filaments are the major region of gas exchange. By restricting gas exchange to such a small surface area, the egg is able to minimize water loss when it finds itself stuck in the drying winds of a dry summer. In addition to a meshwork of air pockets, the respiratory filaments have a water-repellent surface network that maintains a film of gas (a **plastron**) around them when submerged. The plastron allows the respiratory filaments to function as a physical gill if the egg gets trapped in a rain puddle. The egg's chorion, therefore, allows survival of the egg through the rainy season and the dry.

To see some of the complexities of the chorion and respiratory filaments, put a footed coverslip over the egg and adjust your microscope to increase contrast. (A **footed coverslip** is made by nicking the corners of the coverslip against some hard paraffin so that crumbs of paraffin remain attached at each corner. This will give just enough spacer between the slide and the coverslip to avoid crushing the egg. Adjust the size of the crumbs to the size of the eggs. Add more Ringer's solution underneath the coverslip as needed.)

Examine the respiratory filaments under higher power. What differences do you see between them and the rest of the chorion? Include your answers in your notebook with diagrams of what you see. Focus deeper to the level of the embryo within the chorion, and try to stage your embryo using the staging series provided. You will be correcting your stagings after you remove the chorion.

Dechorionating an egg

Soak an egg in undiluted bleach (such as Clorox) for 5 minutes. Agitate the solution slightly with a pipette. You should see remnants of chorion as it separates from the egg. Rinse the egg in Ringer's solution. If the chorion is not gone, repeat the operation.

Note: In moving eggs from one solution to another, a micropipette can be used, but the egg often sticks to the glass and remains within the pipette. Better ways are: to transfer the egg using a hairloop; to catch the egg in the meniscus between half-closed tongs of fine forceps; or to let the egg stick to the side of a needle.

Place the dechorionated egg in a drop of Ringer's solution on a slide, and observe it under a compound microscope. Determine its stage of embryogenesis using the descriptions and staging series below.

Embryogenesis

Study the written description of embryogenesis below, and then proceed with your observations. Look at several embryos to see as many stages of embryogenesis as possible.

Cleavage

The eggs you collect will be in various stages of development. It is difficult to see what is going on inside the egg both because of the chorion (if you haven't removed it) and because the egg is very yolky. It is a **centrolecithal** egg, meaning that the yolk is concentrated centrally and the cytoplasm is pushed to the periphery. Cleavage is unusual in that nuclear division (**karyokinesis**) occurs many times before the cytoplasm cleaves (**cytokinesis**). The nuclei of early cleavage are centrally located through the first seven divisions (Figure 8.2A), after which they start migrating to the periphery (Figure 8.2B). By the time there are about 5000 nuclei all lined up in the peripheral cytoplasm, cell membranes are laid down between them, making a peripheral layer of separate cells, and the embryo goes from being a **syncytial blastoderm** (Figure 8.2C) to a **cellular blastoderm** (Figure 8.2D). This pattern of cleavage is called **superficial** or **peripheral cleavage**.

Cellularization of the blastoderm does not occur simultaneously around the egg. The cells that form first are at the posterior end. They are relatively large and are called the **pole cells** (Figures 8.2C, D). The pole cells form the **primordial germ cells**, which give rise to the gametes. It is interesting that they should be set aside so early in development. Do you have any suggestions as to why?

Gastrulation

Following cleavage, gastrulation proceeds primarily by the infolding of a midventral band of cells. First a **ventral furrow** appears as the mesoderm folds inward (Figure 8.2E). At the anterior and posterior ends of this furrow, the endoderm invaginates forming the anterior and posterior **midgut** (Figure 8.2F). Later the ectoderm will also invaginate to form the anterior **stomodeum** (foregut) and posterior **proctodeum** (hindgut) (Figures 8.2G, H). In addition to these ventral infoldings, there is also a lateral infolding toward the anterior end which extends around the circumference of the egg. This is the **cephalic furrow** (Figure 8.2F) and roughly delineates the boundary of the future head.

Gastrulation creates a multilayered band of germ layers on the ventral side of the egg that curves around the egg's posterior tip. This band is called the **germ band**. It elongates along the dorsal side of the egg so that eventually, like an acrobat with her back arched and her legs gracefully curved back to touch her head, the embryo's posterior end meets its head end (Figures 8.2G, H). The germ band then shortens and thickens, bringing the posterior end of the embryo back toward the posterior pole of the egg (Figures 8.2I, J). As the germ band shortens, definitive **segmental boundaries** appear, marking off head regions (**mandibular, maxillary, labial**), **thoracic segments** (t1–t3), and **abdominal segments** (a1–a10). (In the larva you will find only eight abdominal segments. Abdominal segments 9 and 10 have formed the **telson** of the larva, a tail-like structure.).

During gastrulation, a peculiar thing happens. The developing head disappears from view—it turns inward, or **involutes** (Figure 8.2K). Meanwhile the thoracic segments expand forward, overgrowing the region that used to be "head." Only a tiny external head will remain. So when you finally look at your *Drosophila* larvae, don't be too surprised when they appear to be headless.

Later development

By 16 hours of development, muscular movement will be apparent (Figure 8.2L). Just before the embryo hatches as the first instar larva at 22–24 hours, you will be able to see air-filled **tracheae** and other internal organs.

Embryonic staging series

You can use the staging series shown in Table 8.1 to stage your embryos; it is one of the more widely used series for *Drosophila* embryonic development. (Whenever you refer to a particular stage, you must cite the source of the staging series you've used, since a number of different ones exist.)

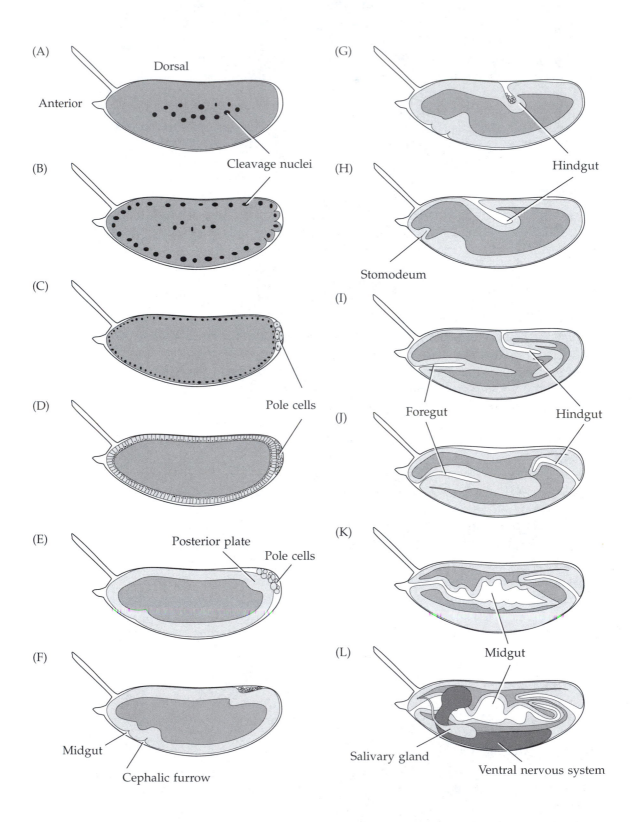

(A)
Anterior
Dorsal
Cleavage nuclei

(B)
Cleavage nuclei

(C)
Pole cells

(D)
Pole cells

(E)
Posterior plate
Pole cells

(F)
Midgut
Cephalic furrow

(G)
Hindgut

(H)
Hindgut
Stomodeum

(I)
Foregut
Hindgut

(J)
Foregut
Hindgut

(K)
Midgut

(L)
Midgut
Salivary gland
Ventral nervous system

◀ **Figure 8.2**
Stages in embryonic development of *Drosophila*. (A) Early cleavage (15 minutes–1.5 hours). Nuclei cleave in the central region. (B) Migration of cleavage nuclei (1.5 hours). Nuclei migrate to the periphery. (C) Formation of syncytial blastoderm (2 hours). Pole cells form posteriorly. (D) Cellular blastoderm (2.5 hours). Cell membranes form between the nuclei. (E) Early gastrulation (3.5 hours). The ventral furrow forms. Thickening of the posterior plate below the pole cells. (F) Midgut invagination (3.5–4 hours). The midgut invagination can be seen ventrally, as can the cephalic furrow. (G) Germ band extension (4–5 hours). Invagination of the hindgut can be seen dorsally. (H) Stomodeal invagination (5–7 hours). Invagination of the stomodeum can be seen ventrally. (I) Shortening of the germ band (9–10 hours). Foregut and hindgut invaginations are deep. (J) Shortened embryo (10–11 hours). Hindgut is now fully posterior. (K) Dorsal closure (13–15 hours). The ectoderm closes dorsally. The midgut broadens. The head involutes. (L) Condensation of ventral nervous system (15 hours–hatching). The gut regions are joined. The nervous system forms ventrally.

Table 8.1 Embryonic stages of *Drosophila*

Stage	Time (hours)	Developmental event (at 25°C)
1	0–0:25	First two nuclear divisions. Egg uniformly dark in center and light at periphery.
2	0:25–1:05	Nuclear divisions 3–8. Egg cytoplasm retracts considerably from vitelline envelope, leaving empty space at anterior and posterior poles.
3	1:05–1:20	At posterior end, three polar buds form (later to pinch off and become pole cells), and divide once. Nuclear division 9. Dividing blastoderm nuclei cause granulated appearance in wide zone in periphery. Empty space at anterior pole disappears.
4	1:20–2:10	Blastoderm nuclei in periphery making a bright peripheral rim. Nuclear divisions 10–13, just prior to cellularization. Two more divisions in polar buds.
5	2:10–2:50	Cellularization of the blastoderm occurs, and nuclei elongate considerably. Pole cells begin to shift dorsally. Midventral blastoderm cells look irregular and wavy, preceding their invagination.
6	2:50–3:00	Early gastrulation: ventral furrow forms, from which mesoderm and endoderm originate; at posterior pole, cells shift dorsally to form a dorsal plate to which pole cells adhere; cephalic furrow becomes visible as a lateroventral slit.
7	3:00–3:10	Endoderm of anterior and posterior midgut and ectoderm of hindgut invaginate; dorsal folds appear.
8	3:10–3:40	Amnioproctodeal invagination, rapid phase of germ band elongation.
9	3:40–4:20	Transient segmentation of the mesodermal layer, visible in region of germ band as prominent bulges protruding into yolk sac.
10	4:20–5:20	Stomodeum invaginates ventrally at anterior pole. Germ band continues to expand. Interior of egg occupied by yolk sac which is a dark, uniform mass. Periodic furrows in epidermis appear. Pole cells leave cavity of posterior midgut and locate themselves dorsally outside yolk sac. Primordia of Malpighian tubules form. Neuroblasts divide.

Stage	Time (hours)	Developmental event (at 25°C)
	Table 8.1 *(continued)*	
11	5:20–7:20	Growth with no major morphogenetic changes. Intersegmental furrows form in epidermis; mandible, maxilla, and labium visible as protuberances. Germ band extension reaches its maximum extent. Posterior pole becomes withdrawn from vitelline envelope.
12	7:20–9:20	Shortening of the germ band so that opening of hindgut becomes located at dorsal side of posterior pole. Width of germ band increases. Anterior and posterior midgut clearly visible and fuse. Germ band segmentation very prominent.
13	9:20–10:20	Germ band shortening completed. Conspicuous triangular gap ventrally due to retraction of clypeolabrum. Labium moves to ventral midline, displacing opening of salivary gland and duct. Yolk sac protrudes dorsally, has characteristic convex shape. Dorsal fold (ridge) appears. Head involution begins.
14	10:20–11:20	Head involution continues. Dorsal closure and closure of midgut. Anal plate ventrally displaced from posterior tip. Dorsal spiracles evident.
15	11:20–13:00	Dorsal closure and dorsal epidermal segmentation. Gut forms closed tube containing yolk sac. Supraoesophageal ganglia and pharynx evident.
16	13:00–16:00	Intersegmental grooves distinguishable middorsally. Dorsal ridge overgrows tip of clypeolabrum. Constrictions appear in midgut. Shortening of ventral cord.
17	16:00–24:00	Tracheal tree contains air. Retraction of ventral cord continues. Embryo hatches as first instar larva.

Source: After Campos-Ortega and Hartenstein, 1985.

Larval Development

You will be maintaining cultures over the week. Place some of your eggs in a small petri dish containing culture medium. Place this dish in a larger petri dish containing sterile water; this will humidify the culture through the week. During the week, keep a record of the developmental stages you see. Record, for example, when you see first instar larvae, second instar, and so on. If one is available, keep a recording thermometer in the room and make a record of the daily temperatures. You may also keep embryos at other temperatures by using the refrigerator (4°C) or an incubator with variable heat settings. (At 22–25°C, development to an adult takes about 9 days. At 10°C, it is slowed down to 57 days. At 29°C, it is speeded up to 8 days. Temperatures continuously above 29°C can be lethal.) By next week, you will have a developmental timetable that you can compare with a standard timetable (Table 8.2). Make comparisons and suggest reasons for any differences you note. Why is development slowed at lower temperatures? Why are high temperatures lethal? Record your answers in your laboratory notebook.

The sections that follow may be done next week.

Table 8.2 Larval stages of *Drosophila*

Time after fertilization		Developmental event (at 25°C)
Hours	**Days**	
24	1	Hatching from egg; first larval instar begins
49	2	First molt; second instar begins
72	3	Second molt; third instar begins
120	5	Puparium formation; puparium white
122	5.1	Puparium fully colored
124	5.2	"Prepupal" molt
132	5.5	Pupation; cephalic complex, wings, legs everted
169	7	Eye pigmentation begins
189	7.9	Bristle pigmentation begins
216	9	Adult ready to emerge from pupa case

Source: After Doane, 1967.

Anatomy of the Larva (Nonsterile Technique Can Be Used)

Put your culture dish or a scoop of medium from the laboratory stocks containing larvae under a dissecting scope. You should see three different sizes of larvae, representing the three larval instars. The first two instars should be found burrowing through the medium. A late third instar will be climbing up, away from the food, getting ready to pupate. Watch the behavior of the larvae as they burrow and eat their way through their food. Make notes in your laboratory notebook. Notice that as the larva feeds, it extends a pair of **mouth hooks** that bring food to the mouth. Look at the head end again. What feature is particularly conspicuous by its absence? Right—the larva has no eyes. Does this mean that the larva is blind? Test its sensitivity to light in various ways, and record your results in your laboratory notebook. You can use a box with a hole at one end to test migration toward or away from light. Don't "toast" your larva—that is, don't place the lamp so close to the larva that you are testing a response to heat rather than a response to light. What other behavioral responses do you note? How does the larva respond when poked, for example? Record all your observations in your laboratory notebook. Note any differences in behavior among the three larval instars.

A few minutes before each larval molt, the larva will stop feeding and lie motionless. The mouth hooks then start biting through the old cuticle. With active muscle contractions, the larva ruptures and leaves its old cuticle, discarding mouthparts and spiracles as well, which are replaced by new structures of the next larval instar. If you are very lucky, or very patient, you may see a larval molt. If you do, time the molt and make whatever observations you can, recording them in your laboratory notebook. (Hint: double mouth hooks and double spiracles [see definition of spiracle below] are evidence of an approaching molt.)

Use a moist paintbrush or partially closed forceps to move larvae from the culture medium to a drop of Ringer's solution on a slide. Try to find all three instar stages. Put the slide over ice in a petri dish to anesthetize the larvae, then observe them again under the dissecting scope.

External anatomy

Look for the tiny external **head**, three **thoracic segments**, eight **abdominal segments**, and a **telson** extending beyond the anus. These segments are delineated by ventral rows of tiny hooks, **denticle belts**, which prevent the larva from slipping backward as it moves forward with waves of extension and contraction. You should also see a number of **sensory bristles** in the first instar larva distributed over the cuticle.

Internal anatomy

The larvae are transparent, and, with the proper lighting, you should be able to distinguish a number of internal structures (Figure 8.3). Use both your dissecting scope and compound microscope, and play with the lighting until you are satisfied. Transmitted light, light coming from below, shining up through the larva, should be best.

The two **fat bodies**—long, whitish sheets running the length of the body—are the most obvious structures. Embedded in the fat body in the fifth abdominal segment are the transparent, vesicular **gonads**. Since the testis is several times larger than the ovary, you can sex the larvae simply on the basis of gonad size.

A tree of beautifully pearl-white **tracheae** (hollow tubes) start anteriorly at the two **anterior spiracles** (tufted openings to the outside) and end in the telson at the two **posterior spiracles**. This is the respiratory system for the larva.

The gut starts anteriorly as a muscular **pharynx** and continues as a narrowed **esophagus**, which runs smack through the middle of the brain in the thoracic region. (There must be a joke or two you can come up with about animals with guts through their brains.) In the region of the esophagus, you should see two lateral transparent **salivary glands**, which you will be dissecting later. The esophagus empties into the heavy-walled, bulblike **proventriculus**, which in turn empties into the **gastric area** that has fingerlike, blind-ending pouches, the **gastric caecae**. The gastric region continues into the long coiled **midgut** or **midintestine**, which doubles back on itself and empties into the straighter **hindgut** or **hind intestine**. You should also be able to distinguish the two yellowish **Malpighian tubules**, the excretory organs that carry urinary waste from the body, emptying it into the posterior midgut.

Focus dorsally and try to see a pulsating blood vessel. This is the **heart**, which extends anteriorly as an **aorta**. The circulatory system is **open**, and the hemolymph bathes the internal organs. Using a clock with a second hand, time the beats of the heart. How does it compare to your own? Does it speed up when the larva is warm? Is there a difference among the three instar larvae? Are there other differences that you note that distinguish the three larval stages from one another? Note these in your laboratory notebook.

Focus again on the region near the brain. Concentrated in this region are the **imaginal discs**. They should appear as tiny teardrop-shaped packets, each one with a connection to the tracheal system. It is these that you will be dissecting. Check Figure 8.3 carefully to determine the position of each type of imaginal disc.

Development of imaginal discs

It is important to understand the development of imaginal discs before dissecting them, so you can appreciate more fully what you will be seeing. The imaginal

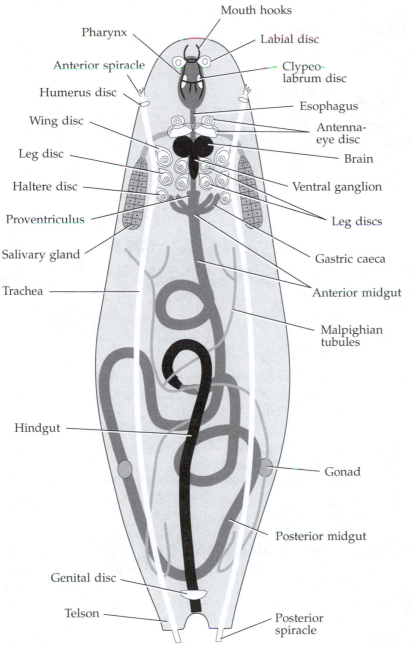

Mouth hooks
Pharynx
Labial disc
Anterior spiracle
Clypeo-
labrum disc
Humerus disc
Esophagus
Wing disc
Antenna-
eye disc
Leg disc
Brain
Haltere disc
Ventral ganglion
Proventriculus
Leg discs
Salivary gland
Gastric caeca
Trachea
Anterior midgut
Malpighian
tubules
Hindgut
Gonad
Posterior midgut
Genital disc
Telson
Posterior
spiracle

Figure 8.3
Schematic diagram of a
third instar larva. The fat
bodies and circulatory
system are not shown for
the sake of clarity.

discs have their origins in the embryo where they start out as epidermal thickenings that then invaginate to become vesicles. They never separate entirely from
the epidermis, but maintain a narrow connection to it. Though there is variation
among the types of discs, the general scheme of development is that once the vesicles form, one side of the vesicle starts growing considerably, bending inward as
it grows, forming a short, wide tube. With nowhere to expand, squashed into the
confines of its little vesicular package, the tube must fold back upon itself as it
grows. The opposite side of the vesicle remains smooth and thin as the **peripodial**

Figure 8.4
Schematic diagram of eversion in a leg imaginal disc. (A) During development of the disc, an epithelial tube forms that becomes folded back upon itself as it is forced into the confines of the tiny disc, bounded by a basement membrane. The epithelium on the far side of the disc does not fold and forms the peripodial membrane. (B) During pupation, the disc everts. The folded epithelium pushes outward against the peripodial membrane, extending into an elongate tube. The distal tip of the tube is shown by five cross lines and the proximal end by small dots in the diagrams of the uneverted and everted discs. (After Poodry and Schneiderman, 1970, and Condic et al., 1991.)

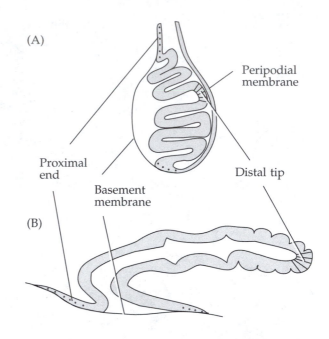

membrane (Figure 8.4A). The original basement membrane, on the outside of the vesicle, does not bend inward with the growing side and remains as an envelope around the imaginal disc. The final effect is a small, transparent package with what looks like a coiled structure (the tube) within. Though not actually coiled, but rather folded like a telescope, this tube will be what forms the final adult structure for which the imaginal disc is determined.

There are nine pairs of discs: **labial**, **clypeolabrum**, **humerus**, **antenna** that is attached to the **eye** disc, **wing**, **haltere**, and three pairs of **leg** discs. There is also a single **genital** disc that will form the genital ducts, accessory glands, and external genitalia, but not the gonads (Figures 8.3 and 8.5). The imaginal discs are suspended within the larval body, looking casually placed, like Christmas ornaments dangling from the tracheal tree. Here they undergo growth but no differentiation.

It is not until the pupal stage, the stage of metamorphosis, that the imaginal discs will take on their adult form. During pupation, each disc everts and its cells **elongate**. The result is that the folded tube within the disc pushes outward against the peripodial membrane, becoming long and extended (Figure 8.4B). (Imagine an old-fashioned top hat with its crown pushed down being snapped to its full height, and you'll have an approximate image of evagination of an imaginal disc.) **Differentiation** of the discs then begins.

Dissection of Imaginal Discs (Nonsterile Technique Can Be Used)

The removal of imaginal discs from the larva requires skill and patience, and it is truly in the realm of microdissection. The imaginal discs you will be dissecting are so small that students before you have dubbed them the "imaginary" discs. Do

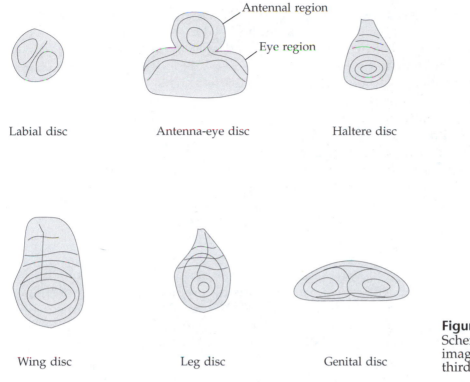

Figure 8.5
Schematic diagram of imaginal discs from a third instar larva.

not despair. Just take a deep breath, don't drink a lot of caffeine before beginning, and be confident in your skills.

Collect late third instar larvae from either your own cultures or the stock cultures kept in the laboratory. Third instar larvae will be the largest and the ones crawling up the sides of the bottles. You could dissect imaginal discs from any of the three instar larval stages, but since the discs have been growing all through the larval stages, they will be largest in the third instar. Rinse one of the larvae in tap water, and transfer it to a drop of insect Ringer's solution on a slide. Place the slide over ice to anesthetize the organism. Your dissection of the discs will be done directly on the slide. Everything but the discs then will be removed from the slide, and a permanent whole-mount preparation will be made of the discs that remain.

Remove your slide from the ice and place it under your dissecting scope. It will be easier to dissect on a lower than higher power (trust me). Use two pairs of fine forceps. With one, hold on to the anterior end of the larva in the region of the mouthparts. Hold on *firmly*. With the second pair, pull on the posterior third of the larva without closing the forceps all the way. A lucky pull will break the body wall just behind the mouthparts and pull it away like a sleeve. The inner parts will be displayed for you in their correct anterior-to-posterior orientation. Use Figure 8.3 to determine what you are looking at. Identify the strands of trachea (tough, white, branched, stringlike structures), the various regions of the gut, the salivary glands, the fat bodies, and the many pairs of imaginal discs hanging to the tracheae. Remember, the discs will look like small, teardrop-shaped packets with a coiled inner structure. (Keep track of the salivary glands; you will be using them later for chromosome squashes.)

Detach the discs using two microneedles. Draw one needle across the other in a scissorlike fashion. Keep manipulating the freed discs to a single, clean spot on your slide until you have collected as many as you can find. Remove the salivary glands to a drop of Ringer's solution on a fresh slide using a microneedle (or pipette, but remember, pipettes have been known to swallow and never relinquish small, soft objects). Invert a petri dish over the slide containing the salivary glands so that they won't dry out before you get to them. Return to your slide of imaginal discs and push the remaining debris off to the side. Remove this debris carefully with a Kimwipe™. Place a drop of 70% alcohol on top of the discs, and record the time in your laboratory notebook. The alcohol will start fixing the discs. Now observe the discs under the dissecting scope, identifying as many as you can (Figure 8.5). Make diagrams of the discs in your laboratory notebook. *Do not let the discs dry out*. Add more 70% alcohol when necessary.

Eversion of imaginal discs

If you have extra leg imaginal discs that you are not saving for whole mounts, and have both time and energy, you can try causing premature eversion of them by soaking them in 0.1% trypsin (an enzyme that acts on proteins) made up in Ringer's solution and adjusted to pH 7.0. (The enzyme solution may be made ahead of time and kept frozen in small alloquots. Unfreeze only the amount you need. The enzyme quickly loses its activity at room temperatures.) You should see eversion within 5–10 minutes. Why do you think this treatment induces eversion?

Whole mount preparations of imaginal discs

It is not easy to mount an entire set of discs. Be satisfied with less. Even a few discs will be adequate. The steps will involve first fixing the discs so they will not deteriorate over time, clearing them so that they are transparent, and finally mounting them in a clear, permanent mounting medium.

You will be using 70% alcohol as a fixative, not because it is the best fixative, but because it will leave no toxic residues in the laboratory. If you want to stain the discs as well, a few drops of toluidine blue can be added to 10 ml of 70% alcohol; this can be used in place of straight 70% alcohol on your discs. The discs should fix for one hour starting from the time you first put alcohol on them (recorded above). *Do not let them dry out* during this period. Put a petri dish cover or some other suitable top over the slide to cut down on evaporation whenever you leave the slide unattended.

You will be clearing the discs in glycerin. This is an excellent clearing agent and is miscible with alcohol and water. Either draw off some of the alcohol surrounding the discs (dangerous) or let some evaporate (much safer) before adding glycerin. Place a drop of glycerin over the discs, then ring the discs with glycerin jelly that has been melted in a warm-water bath. The glycerin jelly is the permanent mounting medium. Now place a footed coverslip over the discs. If there is not enough mounting medium to fill the space under the coverslip, use a pipette to add more glycerin jelly from the side. If there is too much glycerin jelly, leave it until the next lab when it will be thoroughly solid. Then you can clear away the excess by cutting it away with a razor blade.

Kaiser's glycerin jelly

distilled water	52 ml	*Heat to 75°C and stir until dissolved.*
gelatin	8 gm	*Do not heat above this temperature.*
glycerin	50 ml	

1 crystal of thymol (to retard mold)

Observe your slide under the compound microscope, and record what you see with diagrams. You may need to amend your previous identifications, since the imaginal discs now will be much clearer then they were before.

Put your slide face up in a covered box until next week. At that time, use fingernail polish to ring the edge of the coverslip, making an airtight seal. This will prevent the glycerin jelly from drying out.

Chromosome squash from salivary glands

It is common to make chromosome preparations from *Drosophila* larval salivary glands. The reason is that these glands are soft and easily squashed and their cells are large with huge **polytene chromosomes**, chromosomes that replicate without separating. By the third instar, the chromosomes can have as many as 1024 chromatids. Most of the differentiated larval cells are polytene, in fact. Larval cells grow not by mitosis but by duplicating their chromatin and increasing cell size. Only the imaginal discs, histoblasts, and gonadal cells undergo mitosis and remain nonpolytene.

Use the salivary glands on the slide already set aside. Place a drop of aceto-orcein stain on the glands, and cover the slide with a petri dish to avoid evaporation. Stain for 2–5 minutes. Then, without removing any of the stain, place a 22-mm square coverslip over the glands, and with your thumb press down firmly (but not hard enough to break the coverslip). The glands must be completely squashed. Observe under a compound microscope and make a diagram of what you see. A good squash will burst open the cells and splay out the arms of the chromosomes. The banding pattern you see gives a visual map of each chromosome. Geneticists use it to define precisely the location of mapped genes.

Aceto-orcein stain

1% natural orcein in a 1:1 solution of	*Warm gently (do not boil),*
glacial acetic acid and 85% lactic acid	*then cool and filter.*

Pupation

When the third instar larva is ready to pupate, it leaves the medium, its anterior spiracles evert, its body shortens and ceases to move, and it attaches to a firm substrate (such as the side of your bottle). The cuticle then transforms into a **puparium**, which is initially soft and white but soon hardens, turning tan and eventually brown and brittle. Shortly after the puparium forms, the larva detaches from the inside of the puparium by molting a fourth time. Metamorphosis then takes place.

Look at the sides of a culture bottle to see the white-to-brown pupal cases stuck to the side of the glass. These can be released from the glass with a needle

Table 8.3 Stages of metamorphosis in *Drosophila*

Stage	Hours[a]	Developmental event (at 25°C)
P1	0–1	White puparium: wriggling stops completely
P2	1–3	Brown puparium: oral armature stops moving permanently, heart stops pumping, gas bubble becomes visible within abdomen
P3	3–6.5	Bubble prepupa: puparium becomes separated from underlying epidermis; bubble in abdominal region is large, causing prepupa to become positively buoyant at end of this stage (it floats)
P4	6.5–12.5	Buoyant and moving bubble: prepupa is buoyant, and bubble moves, first appearing in the posterior of the puparium, displacing pupa anteriorly, and then appearing in the anterior, displacing the pupa posteriorly. Imaginal head sac is everted and oral armature of larva is expelled
P5	12.5–25	Malpighian tubules migrating and white: legs and wings extend; Malpighian tubules move from thorax to abdomen and become visible as white structures in dorsal anterior abdomen
P6	25–43	Green Malpighian tubules: Malpighian tubules turn green, and dark green "yellow body" appears between the anterior ends of the two Malpighian tubules
P7	43–47	"Yellow body": "yellow body" (actually dark green) moves back between the Malpighian tubules; transparent pupal cuticle separates from underlying epidermis; eye cup becomes yellow at its perimeter
P8	47–57	Yellow-eyed: eyes become bright yellow
P9	57–69	Amber: eyes darken to deep amber
P10	69–73	Red-eye Bald: eyes become bright red; orbital and ocellar bristles and vibrissae darken
P11	73–78	Head and thoracic bristles: head bristles, followed by thoracic bristles, darken
P12	73–78	Wings grey: wings become gray; sex comb darkens
P13	78–87	Wings black: wings become black; tarsal bristles darken and claws become black
P14	87–90	Mature bristles: green patch (the meconium–waste products of pupal metabolism) appears dorsally at posterior tip of abdomen
P15	90–103	Meconium and eclosion: tergites become tan, obscuring Malpighian tubules and "yellow body"; legs twitch; flies able to walk prematurely if puparium removed; eclosion completed

[a] Times start on approximately day four of larval life when the larva is still white but is no longer able to crawl. Timing is variable among individuals, and the times given are a simplification from the Bainbridge and Bownes paper.

Source: After Bainbridge and Bownes, 1981.

or microknife. Examine several under a dissecting scope to see how many stages you can observe.

If time allows, peel a pupa. This can be done by using superglue to glue the pupa to the bottom of a plastic petri dish. When the pupa is secure, use microknives to chip back the pupal case to uncover the tender body of the metamorphosing pupa within. It is well worth the trouble. Remember what is happening during pupation: the larval organs are self-destructing, and the imaginal discs and histoblasts are differentiating to form the adult. To witness this will inspire you, or at the very least change your opinion of these rough brown packages.

Eclosion marks the end of pupation and the beginning of adult life. The insect cracks open the puparium anteriorly and laterally at its seams and emerges from the pupal case. It almost invariably occurs around dawn, when leaves are still damp with dew (*Drosophila* means "lover of dew") and the emerging fly can unfold its new wings and harden its cuticle without the risk of desiccation. The timing of this is controlled by circadian rhythm. If the pupa misses dawn by even a few hours, eclosion will be delayed until the next morning. The dedicated among you will undoubtedly want to rise with the sun and watch this event.

Use Table 8.3 to acquaint yourself with the stages of metamorphosis and stage your pupae.

Accompanying Materials

Tyler, M. S. and R. N. Kozlowski. 2000. *Vade Mecum: An Interactive Guide to Developmental Biology*. Sinauer Associates, Sunderland, MA. "Fruit Fly." This chapter of the CD shows the entire life cycle of *Drosophila melanogaster*, including mating behaviors and ways of sexing the adult and larva. The movies on development include color-codings to indicate germ layers.

Tyler, M. S. and R. N. Kozlowski. 2000. *FlyCycle-II*. Sinauer Associates, Sunderland, MA. This CD-ROM is an adaptation of the 45-minute film, *Fly Cycle: The Lives of a Fly*, Drosophila melanogaster, 1996 by M.S. Tyler, J. W. Schnetzer and D. Tartaglia, Sinauer Associates, Sunderland, MA. This covers the life cycle of the fruit fly as well as a number of the mutants used in research.

Gilbert, S. F. 2000. *Developmental Biology*, 6th Ed. Sinauer Associates, Sunderland, MA. Chapters 9, 18, and 19. In these chapters you will find an excellent discussion of *Drosophila* development, larval polytene chromosomes, imaginal discs, and metamorphosis. The diagrams and photographs throughout are extremely useful.

Fink, R. (ed.). 1991. *A Dozen Eggs: Time Lapse-Microscopy of Normal Development*. Sinauer Associates, Sunderland, MA. Sequence 6. This shows *Drosophila* embryogenesis from cleavage to hatching.

Selected Bibliography

Ashburner, M. et al. (eds.). 1976–1986. *The Genetics and Biology of* Drosophila. Academic Press, New York. This is a series of volumes of collected papers. Sophisticated and technical, it is well worth browsing through. Volumes 2a–2e, edited by Ashburner and T. R. F. Wright, concentrate on developmental and biochemical studies.

Bainbridge, S. P. and M. Bownes. 1981. Staging the metamorphosis of *Drosophila melanogaster. J. Embryol. Exp. Morph.* 66: 57–80. This well-illustrated paper gives detailed descriptions of each stage of metamorphosis.

Campos-Ortega, J.A. and V. Hartenstein. 1985. *The Embryonic Development of* Drosophila melanogaster. Springer-Verlag, Berlin. This is the authoritative reference on *Drosophila* development. It is well illustrated and explains each stage of development.

Condic, M. L., D. Fristrom and J. W. Fristrom. 1991. Apical cell shape changes during *Drosophila* imaginal disc elongation: A novel morphogenetic mechanism. *Development* 111: 23–33. Though dealing primarily with the details of cell-shape changes, it also provides a good start on a bibliography for imaginal discs.

Demerec, M. (ed.). 1994. *Biology of* Drosophila. Cold Spring Harbor Laboratory Press, New York. This is a facsimile edition of the original 1950 publication. It is the *Drosophila* bible, an excellent classic, and a comprehensive guide to the histology and development of all stages in the *Drosophila* life cycle.

Demerec, M. and B. P. Kaufmann. 1986. Drosophila *Guide.* Carnegie Institute of Washington, Washington, D.C. A very inexpensive, short paperback guide to the life cycle, breeding methods, and genetic techniques for *Drosophila.*

Doane, W. W. 1967. *Drosophila.* In *Methods in Developmental Biology,* F. H. Wilt and N. K. Wessells (eds.). Thomas Y. Crowell Co., New York, pp. 219–244. The entire text is superb, covering a number of different species and describing rearing and experimental methods for them.

Hall, J. C. 1994. The mating of a fly. *Science* 264: 1702–1715. A lengthy review of courtship and mating in *Drosophila,* including details on genetics and molecular biology.

Hipfner, D. R. and S. M. Cohen. 1999. New growth factors for imaginal discs. *BioEssays* 21: 718–720. A brief and excellent review of how imaginal discs grow during the larval stages.

Lawrence, P. A. 1992. *The Making of a Fly: The Genetics of Animal Design.* Blackwell Scientific, Oxford. This book is beautifully illustrated, making it useful beyond its genetic emphasis.

Leptin, M. 1994. *Drosophila.* In *Embryos, Color Atlas of Development,* J. B. L. Bard (ed.). Wolfe Publ., London, pp. 113–134. A well-written review of *Drosophila* development that is beautifully illustrated.

Poodry, C. A. and H. A. Schneiderman. 1970. The ultrastructure of the developing leg of *Drosophila melanogaster. Wilhelm Roux Arch.* 166: 1–44. This is an excellent paper on the eversion process in a leg imaginal disc.

Roberts, D. B. (ed.). 1986. Drosophila: *A Practical Approach.* IRL Press, Oxford. This is a gold mine of well-explained techniques and general information. One chapter, "Looking at Embryos," is particularly useful for this laboratory study.

Spieth, H. T. and J. M. Ringo. 1983. Mating behavior and sexual isolation in *Drosophila.* In *Genetics and Biology of* Drosophila, Vol. 3c, M. Ashburner, H. L. Carlson and J. N. Thompson, Jr. (eds.),. Academic Press, New York, pp. 223–284. This very thorough review describes each mating behavior, discusses variations among species, includes pictures of courtship behavior, and analyzes its adaptive significance.

Treisman, J. E. 1999. A conserved blueprint for the eye? *BioEssays* 21: 843–850. A clear review of the recent evidence that the genetics of eye development in *Drosophila* and vertebrates share a number of common features.

Wilkins, A. S. 1986. *Genetic Analysis of Animal Development*. Wiley-Interscience, New York. An excellent book with several descriptive chapters on the embryonic and larval development in *Drosophila*.

Suppliers

Most biology or zoology departments have someone with a supply of *Drosophila* cultures "in the cupboard" (which they must constantly clean out to keep active). A kind word can often get you all the *Drosophila* material that you need for this lab. If the source is not in-house, however, a number of supply companies provide cultures.

Nasco

901 Janesville Ave.
Fort Atkinson, WI 53538-0901
1-800-558-9595
www.nascofa.com

Sells inexpensive *Drosophila* cultures, wild-type and mutants. Also sells *Drosophila* medium, though it is certainly easy enough to make your own.

Connecticut Valley Biological Supply Co., Inc.

P.O. Box 326
82 Valley Road
Southampton, MA 01073
1-800-628-7748.

Sells stock cultures including eggs, larvae, pupae, and adults. Must order at least three weeks in advance. Also sells a number of mutant strains.

Any good chemical supply company such as:

Sigma Chemical Company

P.O. Box 14508
St. Louis, MO 63178-9916
1-800-325-3010
www.sigma-aldrich.com

brewer's yeast (also from natural food stores)
glacial acetic acid
glycerin
lactic acid
natural orcein
propionic acid
salts for Ringer's solution
thymol

chapter 9 Early Development of the Chick
Gallus domesticus

The common domestic chick, *Gallus domesticus*, is traditionally studied in embryology classes. Its development is an international language common to all students of embryology, and this alone is reason enough to learn about it. There are other reasons as well. Living chick embryos are easily obtained. Since chickens breed in any season, chick embryos often are no further away than the nearest farm. Above all, chick embryos can teach you a great deal about vertebrate development in general and will help you understand comparative vertebrate anatomy and vertebrate evolution. You will see the vertebrate body plan being laid down, learn how the organs form, and witness how they relate developmentally to one another. You will see the structures form that are diagnostic characters of the phylum Chordata and subphylum Vertebrata. And you can extrapolate from chick to human development and suddenly know a tremendous amount about what must go right for a normal baby to be born and what can go wrong that causes birth defects. Learn chick developmental anatomy well, and I promise it will serve you well.

Throughout the chapter, you will find a number of questions (easy ones). Remember to record answers to these questions in your laboratory notebook.

The Chick Egg

The female chicken lays an **amniote egg**, the same type of egg laid by other birds, reptiles, and monotreme mammals. An evolutionary invention of the early reptiles, it is one of the major changes that freed vertebrates from the water, allowing them to lay eggs on land. The design is an engineering masterpiece. An outer shell protects against physical damage and dehydration while still allowing for gas exchange. Packed within is enough food and water to survive the long journey to hatching—and a long journey it is. In the chick, it takes 21 days. Stranded out on land, the embryo cannot rely on an aquatic feeding larval stage, as do, for example, most amphibian embryos. The provisions packed within the shell must be enough to nourish all of development through to the hatching of the juvenile. In a medium-size chick egg (about 60 gm), these provisions include more protein (7.2 gm) than there is in a frankfurter; as much fat (6 gm) as is in 2 teaspoons of butter but with about seven times the cholesterol (300 mg); and about 8.5 teaspoons of water (40 gm). Despite the bad reputation of its cholesterol, the egg truly is an excellent food source.

In order to permanently alter the way you look at an egg in the kitchen, you will be dissecting one in the laboratory. Obtain an egg that was collected from the wilds of a local grocery store. Examine the shell and see that one end is blunt. This marks the region of an air space that you will see upon opening the egg. Prior to opening your egg, put it briefly into hot water (water that has been brought to a boil, but is no longer boiling), or watch someone else do this with their egg. What happens? What does this tell you?

Now open your egg *very carefully* into a finger bowl without breaking the yolk. Do this by giving the egg a sharp crack against the edge of the bowl, putting your two thumbs in the crack, and pulling the two halves apart, with the egg close to the bottom of the finger bowl. If you are lucky, you have kept the chick ovum (the yolk) intact. No wonder flipping a fried egg without bursting the yolk is an art—that yolk is one cell. It's huge! Only a fragile **cell membrane** overlain by a **vitelline envelope** protects it from the slice of your spatula. Compare the yolk to a more typically sized cell—a piece of dandruff, for example, which represents a single epidermal cell. The difference is remarkable! Most of the bulk in the chick ovum is not cytoplasm, but spheres of yolk lipid suspended in a sea of yolk protein. The yolk is so disproportionately abundant compared to the cytoplasm that the cytoplasm is displaced to the animal pole and sits as a puddle on top of the yolk. The large quantity of yolk classifies the egg as **macrolecithal** (*macro*, large; *lecithal*, yolk).

Examine the eggshell. Ninety-eight percent of it is **calcite**, which is calcium carbonate. Calcium is the major component of bone as well (in the form of calcium phosphate, or apatite). The calcium that goes into the shell comes both from the hen's diet and from her bones. Weigh the shell. Realize that a truly remarkable hen can lay an egg a day. Approximately how much calcium will a one-a-day hen lose in a week? In a year, a hen can put into her eggshells over 25 times her own skeletal weight in calcium. Obviously, this must be replaced. The standard feed given to laying hens is 3 to 4% calcium, and one hen receives about a 100 gm of feed a day. Based on the weight of the shell, is this sufficient to replace the calcium lost in a superhen laying every day?

Examine a piece of shell under a dissecting scope, turning it so that you can see it in cross section. Notice that it has three layers (Figure 9.1): two calcified layers,

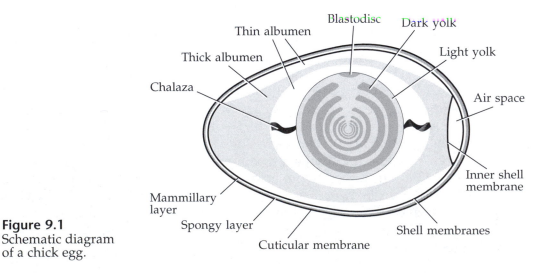

Figure 9.1
Schematic diagram
of a chick egg.

the inner **mammillary layer** and the outer **spongy** (or **crystalline**) **layer**; and outermost the thin shiny **cuticular membrane**. In the mineralized layers, calcium carbonate is laid down as crystals perpendicular to the surface of the egg—this gives the shell tremendous strength. The outer cuticular membrane is made of glycoprotein. Prior to laying, it is wet and slippery, which helps with oviposition, but soon after laying it dries out and serves to protect against invasion by microorganisms.

Notice the shell has numerous **pores**. Use food coloring to dye the shell. When the dye dries, a greater amount will have collected in the dips of the pores, and these will appear as dark spots. Notice that the dark spots are irregularly sized and spaced, and that they are more abundant in the area over the air space than anywhere else. Can air pass through these pores? How can you tell? Think back to what happened when you put the egg into hot water. If air could not pass through the shell, what would happen to the embryo? (Write your answers in your laboratory notebook.) Wherever air can pass, water vapor will follow. Balancing the need for gas exchange against the risks of dehydration while retaining strength is always a major task in shell design.

On the inside of the shell you should see two **shell membranes**, an inner and outer one, separated from one another at the region of the air space (**air cell**) at the blunt end of the egg (Figure 9.1). Peel the two membranes apart from one another. Which of the two membranes is thicker? Look at a piece of each membrane under the dissecting scope, then stain them with 0.1% toluidine blue and mount each on a slide with a coverslip using water as mounting medium. You will see that the membranes are a dense, crisscrossing mat of fibers. These elastin-like fibers retard water loss, give mechanical support, and provide a defense against microbes. Where else do you find a high proportion of elastin? (In ligaments: attaching bone to bone, they must be able to s-t-r-e-t-c-h; in the cartilage of the external ear: pull on your ear—does it droop, or bounce back into shape?) What does this tell you about the mechanical properties of the shell membranes? Experiment to see how much stretch the membranes have.

Look at the air space at the blunt end of the egg. This space forms between the two shell membranes after the egg is laid. As the egg cools from the internal temperature of the hen (about 41°C) to room temperature, the internal contents contract more than the shell, creating the air space. Try cooling a warm egg by putting it into the refrigerator, and see if the air space becomes even larger. Do you think the air space serves a function? Could it help to absorb shock when the egg is jarred? When the chick begins to hatch, it first breaks through into the air space using its beak and fills its lungs with air. The confined space soon becomes fouled with carbon dioxide, which triggers the chick to start breaking through its shell (a process referred to as **pipping**, a word most of us have heard without knowing its meaning. (In the folk song "I Gave My Love a Cherry," one verse goes, "A chicken when it's pipping, it has no bone." Does this make any sense?)

Now look at the egg **albumen**. Notice that it has varying viscosities. Around the yolk is a narrow band of thin albumen, surrounded by a layer of thick albumen, and external to this another layer of thin albumen (Figure 9.1). The albumen is 88% water and is a major source of water for the developing embryo. The rest is primarily glycoproteins, the most abundant of which is **ovalbumen**. Some of the proteins in the albumen serve antibacterial functions: **lysozyme**, for example, is an enzyme that disrupts bacterial cell walls; **ovotransferrin** binds iron and **avidin** binds the vitamin biotin, making these nutrients—which are necessary for bacterial growth—unavailable to the bacteria. Two other proteins, **ovomucin** and **cystatin**, are thought to have antiviral activity. In addition, the albumen has a pH that prevents bacterial growth. What pH is this? Measure the pH using pH paper. (Did your parents, when making lemon meringue pie, ever tell you not to lick the meringue bowl because of the raw egg whites? What they were worried about was the avidin, which can cause a vitamin deficiency called egg white disease. Explain. Why are cooked egg whites all right to eat?)

Notice that part of the albumen is very dense and is coiled into two cordlike structures attached tightly to the yolk membrane. These are the **chalazae** (pronounced *ka-lay-zee*) (Figure 9.1). They suspend the yolk in the middle of the albumen and allow it to rotate. The yolk will rotate, always orienting to gravity so that its animal pole (and the embryo) face upward. Why do you think this is important? (By definition the animal pole of an egg is the pole where the polar bodies form. In the chick, this is also where the blastodisc is. The vegetal pole is opposite the animal pole.) Remove the chalazae using scissors and fine forceps and place them side by side on a slide, keeping track of which end was attached to the yolk. Do the fibers twist the same way? Add a drop of 0.1% toluidine blue stain to help distinguish the fibers. You probably noticed that one chalaza twists clockwise and the other counterclockwise. This twisting occurs during egg formation as the yolk is slowly rotated during its descent through the oviduct. This twists the fibers of the chalazae and draws them taut, pulling the yolk to the center of the egg. You can visualize this using a cork, two rubber bands, and thumbtacks. Attach the rubber bands with thumbtacks to opposite ends of the cork. Holding the free ends of the rubber bands, slowly rotate the cork as the yolk would rotate coming down the oviduct. Notice that the rubber bands twist in opposite directions, just as the chalazae do. Notice also that as the slack in the rubber bands is taken up, the cork is suspended between the two ends. Unlike rubber bands, however, the chalazae do not tend to unwind. Using microneedles, try to tease apart the fibers of the chalazae and notice how resistant they are to untwisting.

Examine the **yolk**. Where does the yolk sit in relation to the albumen? Does it float, or is it submerged? Remember that bowl full of eggs before you made your last omelet. Did the yolks float or sink in that sea of whites? What does this tell you about the relative densities of yolk and albumen? About the importance of the chalazae? Notice the **blastodisc**, the puddle of cytoplasm that should be oriented upward. It is here that the egg nucleus is found. With forceps and fine scissors, remove the blastodisc by puncturing the vitelline envelope and cell membrane with the forceps, and holding on to both with the forceps while cutting around the blastodisc with the scissors. Carefully lift the blastodisc to a slide. A disc of white yolk will be attached to the underside of the cytoplasm. Examine the blastodisc under the dissecting and compound microscopes. Remember that the nucleus is partway through a meiotic division, so there is no nuclear membrane, and the chromosomes are aligned on a metaphase plate; they are very difficult to detect.

If you see a spot of blood in the yolk, this can be either an indication of a developing embryo (it would have to be in the location of the blastodisc—look at it carefully) or simply the incorporation of a blood clot from the ovary at the time of ovulation.

The yellow color of the yolk is due to **carotenoid** pigments—the same pigments seen in carrots and other yellow vegetables. (Your body uses carotenoids to make vitamin A, which has a long list of functions. When transformed into retinal, for example, it becomes part of an eye pigment, rhodopsin, necessary for vision; when transformed into retinol, it is needed for skin maintenance. Egg yolk has about one-ninth as much carotinoid as carrots.) The carotinoid pigments are absorbed from the hen's food and deposited rapidly in the yolk. Consequently, greater amounts are deposited during the day when the hen is eating than at night when the hen is sleeping. This results in the yolk having concentric layers of dark and light yolk (Figure 9.1), and provides us with a daily record of the yolk deposits. You can see this by removing an unbroken yolk with a spoon to some boiling water and boiling it for 10 minutes. Then, using a razor blade or sharp scalpel, carefully bisect the yolk through the middle, starting at the animal pole where the blastodisc is. Count the rings. Each sequence of a light and dark ring equals one day. How many days did it take the hen to form your yolk? Remember when and where the yolk was being packed into the ovum (in the ovary prior to ovulation).

The Female Chick Reproductive Tract

The female chick has only one fully developed ovary and oviduct (usually the left), while the other remains vestigial. Why do you think this is so? If your laboratory has a female chick reproductive tract on display, examine this, using Figure 9.2 to study each region discussed below.

Look at the **ovary**. Those large lumps are oocytes developing to maturity. These are ovulated as secondary oocytes, when the egg is in metaphase of its second meiotic division. Meiosis is not completed until after fertilization. Fertilization takes place high in the oviduct, in the region called the infundibulum. Why

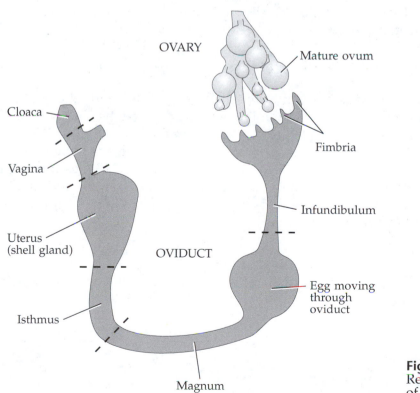

Figure 9.2
Reproductive tract of a female chick.

would it be difficult for fertilization to occur farther down in the oviduct? After ovulation, the egg remains fertilizable for only about 40 minutes. This does not mean, however, that mating must have occurred near the time of ovulation. Sperm can be stored in the female tract and remain viable there for as long as 35 days, though viability is reduced after the third week. Look back at the slides of sperm from an earlier laboratory session to recall the shape of chicken sperm. At the time of fertilization, many sperm enter the egg, but only one sperm pronucleus fuses with the egg pronucleus. This unusual phenomenon is called **physiological polyspermy**.

Notice that the mouth of the **oviduct** is not attached to the ovary. It is not a given that every egg that is ovulated successfully enters the oviduct. In fact, as many as 5% of eggs ovulated normally miss the opening and fall into the body cavity, where usually they are absorbed. (A hen that does this a high percentage of the time is called an "internal layer.") Once in the oviduct, the ovum packed with yolk is moved along by muscular contractions of the oviduct. After passing through the infundibulum, albumen begins to be secreted around the yolk. Most of the albumen is secreted by cells in the region of the oviduct called the **magnum**. The egg passes from the magnum through the **isthmus**, where the chalazae and the inner and outer shell membranes are added. In the **uterus**, also called the **shell gland**, water and salts traverse the shell membranes and are added to the albumen, **plumping** out the shell membranes. The shell is then deposited, and the egg travels on through the vagina and into the **cloaca**. Here the egg is turned completely around. Having traveled with its small end first, its orientation is now reversed, and the egg is laid with the large blunt end first. The entire process from ovulation to laying takes about 24 hours, the major portion of which (about 19 hours) is spent in the shell gland. If the egg was fertilized, development of course has been taking place all this time, and the embryo is already undergoing gastrulation by the time it is laid.

Cleavage

Cleavage stages cannot be observed in the living egg under normal conditions, since they occur while the egg is still in the oviduct. You will be studying cleavage, therefore, by examining models and diagrams of the events (Figures 9.3 and 9.4). Remember the function of cleavage. The fertilized egg already contains within its cytoplasm all of the germ-layer components. Cleavage divides this single cell into a lot of smaller building blocks that then can be rearranged and molded into the multicellular organism. Use the models and diagrams to identify all of the structures printed in boldface in the next two paragraphs.

Figure 9.3
Early cleavage stages in the chick. The blastoderm is shown removed from the yolk and viewed from above, starting with the 2-cell stage and showing progressively more advanced stages.

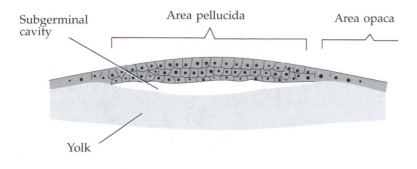

Figure 9.4
Late cleavage stages in the chick. The animal pole of the egg is shown in cross section. This stage occurs within the oviduct, 3–4 hours prior to laying.

The first thing you should notice is that the entire egg does not cleave; only the puddle of cytoplasm sitting on top of the yolk cleaves. Dense yolk creates a formidable impediment to a cleavage furrow. You will see this later in the mesolecithal egg of the amphibian, in which the moderate amount of yolk concentrated in the vegetal half slows down the cleavage furrows. Here in the chick egg, however, the huge amount of yolk stops the cleavage furrows altogether. Cleavage initially, therefore, is necessarily partial, or **meroblastic**. Cleavage furrows, which cut down from the animal pole toward the yolk, stop when they hit the yolk. The cleavage pattern in chick is also often called **discoidal**, since it is restricted to the circular disc of cytoplasm, the blastodisc.

The cells produced during cleavage are called **blastomeres**, and together they make up what is called the **blastoderm**. In the center of the blastoderm, the cells become separated from the underlying yolk by a space called the **subgerminal cavity**. The space makes the central area of blastoderm look lighter and more translucent than the surrounding area. This central region is called the **area pellucida** (a *pellucid* stream is transparent; a *pellucid* explanation is clear and easily understood). The surrounding area of blastoderm, still connected to the underlying yolk, looks darker and opaque; it is called the **area opaca**.

Gastrulation

By the time the egg is laid, the embryo is undergoing gastrulation. The cell movements during this stage have been traced by scientists working with great patience and steady hands. The normal procedure is to label tiny regions of the blastodisc with vital dyes and other markers, to watch the paths of movement these markers make during gastrulation, and to determine where they eventually end up, thereby determining the final fate of the cells. In this way, maps have been constructed showing the location of each germ layer prior to gastrulation and its route of travel during gastrulation. These maps, called **fate maps**, are a gift to any experimenter, providing a blueprint for a vast array of experiments. As you study them, think of experiments you would do using the fate maps as a ground plan. Briefly outline one of your experiments in your laboratory notebook.

To begin your study of gastrulation, first remember what gastrulation achieves. It separates the germ layers so that they are in appropriate positions for the organ formation that will follow. Since endoderm forms the epithelium of the gut and gut derivatives, it must become a centrally located tube. The ectoderm, which forms the epidermis and nervous system, must cover the outside. And in between must come all the packing material provided by the mesoderm—muscle, skeleton, circulatory system, kidney, and dermis. Now look at the fate map of the chick blasto-derm (Figure 9.5). Notice that it is laid out flat as a pancake, a configuration forced upon it by the huge inert yolk. This is a problem. Embryos such as those of sea urchins and amphibians don't face this problem. Because they form a hollow sphere as blastulae, they are able to gastrulate by simply moving the mesoderm and endoderm inside the sphere and then arranging these germ layers appropri-ately within the sphere. But in the chick, what is a sphere in these other organisms is laid out flat, and the germ layers are forced to separate in a flat plane forming a three-layered sandwich with endoderm on the bottom, ectoderm on top, and mesoderm in between. The flat germ layers later get tucked and folded to form the gut tube and other structures. Meanwhile, the entire sandwich is growing outward over the surface of the yolk and eventually surrounds the yolk.

It is only the central region of the gastrula sandwich that will form the body of the embryo. The outer regions will form extraembryonic membranes: the **amnion**, **chorion**, **allantois**, and **yolk sac**. (The amnion and chorion serve to protect the em-bryo; the allantois stores nitrogenous waste and where it is attached to the chori-on provides a vast respiratory surface that also resorbs calcium from the eggshell for the growing embryo; the yolk sac absorbs yolk and transports it back to the embryo in its vitelline blood vessels. You will see these in later chapters.)

Look at the blastoderm fate map again, and imagine how you would orchestrate the gastrulation movements to create a three-tiered sandwich. Now look at Figures 9.6 and 9.7 and at the models of chick gastrulation to see how close your plan of gas-trulation came to what actually happens. The first step in chick gastrulation is the splitting off of the major bulk of endoderm from the underside of the blastoderm to form a layer of endoderm below the rest of the cells. This splitting is called delami-nation. What molecular mechanisms must be involved in **delamination**? Obviously, the endoderm must lose its stickiness to the rest of the blastoderm; it must change its affinity by changing its cell surface adhesion molecules. If you were able to con-duct experiments, how might you test for this?

Delamination produces an upper epiblast, consisting of the presumptive ecto-derm, mesoderm, and some endoderm; and a lower **hypoblast**, consisting of the **extraembryonic endoderm**. The extraembryonic endoderm will participate in

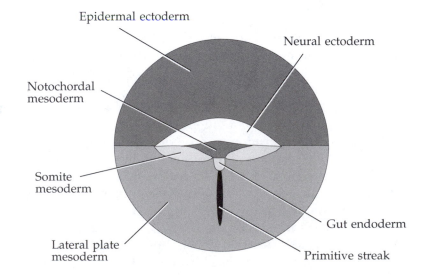

Figure 9.5
Fate map of the chick gastrula. The blasto-derm, at about the time of laying, is shown from above so that only the epiblast is seen. The extraem-bryonic endoderm has already delami-nated to form the hy-poblast and can not be seen in this view.

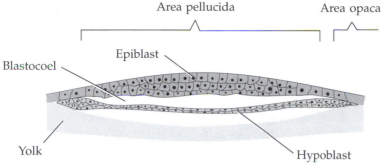

Figure 9.6
Early gastrula stage in the chick. The animal pole is shown in cross section. The extraembryonic endoderm has delaminated to form the hypoblast lying below the epiblast. Compare with Figure 9.4.

forming extraembryonic membranes. The presumptive endoderm remaining in the epiblast, the **embryonic endoderm**, will form the endodermal derivatives of the embryo itself, but it must first join the hypoblast before doing so.

The embryonic endodermal cells move to the hypoblast by turning inward along a line of ingression and then moving downward to the hypoblast. The line along which cells are moving inward from the epiblast, or ingressing, is called the **primitive streak**. The prospective **mesoderm** cells are the next to enter the primitive streak and leave the epiblast. However, they do not move down to join the hypoblast but instead make a U-turn and migrate outward between the epiblast and hypoblast to form a separate layer between the two.

Compare this type of gastrulation to that of a sea urchin or amphibian. What in the sea urchin or amphibian would be analogous to the primitive streak? Yes, the blastopore. Remember the dorsal lip of the blastopore in an amphibian gastrula? It is where notochordal mesoderm cells involute to form the notochord. There is an analogous structure in the chick embryo at the top of the primitive streak called **Hensen's node** (or **primitive node**). It is here that the **notochordal** cells converge, ingress, and migrate forward, rather than laterally, to form a streak of notochordal cells down the midline.

Look at various stages of gastrulation, noticing that gastrulation begins in anterior regions first and progresses to more and more posterior regions. Anterior regions may have finished gastrulating and gone on to organ formation while

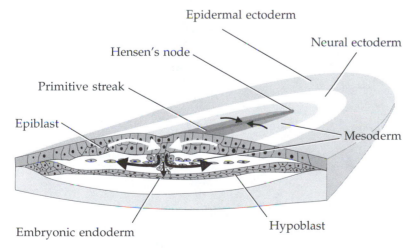

Figure 9.7
Mid-gastrula stage in the chick. A three-dimensional view of the anterior half of a gastrulating chick shows the direction of movement of the cells. Embryonic endoderm and mesoderm are ingressing through the primitive streak. The endoderm is migrating down to join the hypoblast, and the mesoderm is migrating outward between the epiblast and hypoblast forming an intermediate layer.

posterior regions are still gastrulating. This anterior-to-posterior wave of progression results in the primitive streak becoming confined to ever-more posterior regions. This is often referred to as the **regression of the primitive streak**.

24-Hour Chick Whole Mount

Molding a body out of the three-tiered sandwich of a chick gastrula is a task that takes a lot of folding and pinching. The beginnings of this process are well illustrated in a 24-hour chick embryo (by tradition, this means an embryo incubated for 24 hours—total time of development has been more than 24 hours, since development began prior to egg laying) (Figure 9.8). You will be looking at whole mounts in which the cellular region has been removed from the yolk and mounted on a slide. Because of the transparency of the embryo at this stage, all levels can be seen by focusing up and down through the embryo. Use both your dissecting and compound microscopes, experimenting with different settings for contrast and lighting. Be able to identify any of the structures printed in boldface below, and write in your laboratory notebook answers to any questions posed. Realize that you yourself looked very much like this embryo at one time (about three weeks into your development).

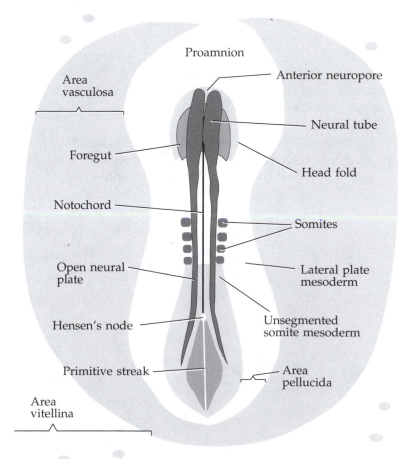

Figure 9.8
Schematic diagram of a 24-hour chick embryo.

By 24 hours of incubation, the anterior half of the embryo is undergoing neurulation while the posterior half is still gastrulating. Notice that the **area opaca** has greatly expanded. In fact, it now occupies too large of an area to be mounted on your slide, so much of it already has been trimmed away. Even so, you still should be able to identify the two subdivisions of the area opaca: the outer **area vitellina** (*vitellina* means yolk), where cells contain many yolk granules; and the inner **area vasculosa** (referring to vascularization), where **blood islands** are forming. Blood islands are masses of blood-forming cells that will later anastomose to form a capillary network that will bring yolk nutrients back to the embryo proper. Look at the **area pellucida**, the clear area around the embryo; it looks almost like a footprint. Just in front of the embryo's head is a region of the area pellucida that is particularly clear. This is called the **proamnion** (an unfortunate name whose meaning, "before the amnion," has nothing to do with the structure—suggest a better name and we'll vote on it). The reason the proamnion is so translucent is that it consists only of ectoderm and endoderm. The spreading mesoderm has not yet reached this area.

Focus now on the embryonal area. The folding and pinching has already begun. At the anterior end you will see a **head fold**. This is where an anterior fold has undercut the developing head and raised it above the level of the blastoderm. Look at the models on display to see this more clearly. Notice how the head fold has caused the endoderm to fold into a closed tube to form the **foregut**. Posteriorly, where the endoderm is still flat, is the open **midgut**, and the transition between the closed foregut and open midgut is the **anterior intestinal portal** (*portal* means door). This is best visualized by looking at models. Now look carefully on your whole mount for the darker outline of the foregut and the crescent-shaped line of the anterior intestinal portal. Look at models to see the mesoderm lying directly underneath the foregut and lateral to it. This mesoderm will be forming the heart later on and is therefore called the **cardiac mesoderm**. On your whole mount, it should look somewhat darker than the surrounding region.

One of the most notable features of the 24-hour embryo is the invaginating **neural ectoderm**. Previously it was an open **neural plate**, but by 24 hours, the anterior region has folded upward to form a closed **neural tube**. A major part of this folding is caused simply by a change in shape of the neural plate cells from cuboidal to truncated-pyrimidal. Convince yourself of this by making a sheet of cells (neural plate) out of small cubes of clay. Now change the shape of each cube by pinching its apical end (a contraction due to an apical belt of actin and myosin microfilaments in the cells of the neural plate) and elongating it along its apical-basal axis (caused by an elongation of microtubules in this axis). Now fit the clay cells back together, and you will see that your sheet of cells has folded upward.

Look at your whole mount again. Notice that neurulation is taking place earlier anteriorly than posteriorly. Identify regions of **closed neural tube** (if any), **open neural plate**, and in between regions where **neural folds** have formed but have not closed. Notice that at the anterior tip of the embryo, the neural plate does not close completely but leaves an open channel called the **anterior neuropore**. This will be closed by 36 hours. (Sometimes the neural folds in the spinal region fail to close during development, causing an extremely severe birth defect called spina bifida with myeloschisis. Occurring in humans as well as chicks, vertebrae do not form properly in the area, and the spinal cord is left an unprotected, flattened mass of neural tissue.)

One thing you cannot see in the whole mount, which you will see later in serial sections, are the neural crest cells lying on top of the neural tube. As soon as the tube closes, they start migrating away from the neural tube, sometimes to distant regions. These are fascinating cells and will be discussed more completely in the next chapter. For now, just keep in mind that they are unique to the vertebrates and form a variety of structures including cartilage and membrane bones of the head, pigment cells of the skin, and spinal ganglia.

Look at the open midgut region, and notice that the mesoderm has segmented into blocks of **somites** on either side of the **neural tube**. Count and record the number of somites your embryo has. As the embryo develops, additional somites will form from the more posterior **unsegmented somite mesoderm**. Recall what the somites will be forming: the axial skeleton (vertebrae and ribs), muscles of the axial skeleton and limbs, and dermis. Between the two rows of somites, in the midline underneath the neural ectoderm, you should see a streak of condensed tissue. This is the **notochord**. How far does it extend anteriorly and posteriorly on your embryo?

At the posterior end of the embryo, notice the region of the **primitive streak** and **Hensen's node**, showing that ingression is still occurring in this region. Can you think of any reason why an embryo would spend more time making its anterior end than its posterior end?

Now that you are thoroughly familiar with your embryo, you can stage it more accurately than simply calling it a 24-hour embryo. You can use the staging series developed by Hamburger and Hamilton; a truncated version of this series is given on pages 127–128 in the next chapter.) This is the most widely used staging series for the chick. Record the stage of your chick. By convention, when using this staging series, you would precede the stage number with the initials HH.

Accompanying Materials

Tyler, M. S. and R. N. Kozlowski. 2000. *Vade Mecum: An Interactive Guide to Developmental Biology*. Sinauer Associates, Sunderland, MA. "Chick Early." This chapter of the CD shows step by step how to open the egg and identify its parts. It shows early development in a series of movies developed from models and histological whole mounts.

Gilbert, S. F. 2000. *Developmental Biology*, 6th Ed. Sinauer Associates, Sunderland, MA. Chapters 2 and 11. In these chapters are clear descriptions and illustrations of the amniote egg, chick cleavage and gastrulation, and a discussion of the mechanisms of the gastrulation movements.

Fink, R. (ed.). 1991. *A Dozen Eggs: Time-Lapse Microscopy of Normal Development*. Sinauer Associates, Sunderland, MA. Sequence 11. This shows early chick development, allowing you to see neurulation and regression of the primitive streak.

Selected Bibliography

Burley, R. W. and D. V. Vadehra. 1989. *The Avian Egg: Chemistry and Biology*. John Wiley and Sons, New York. Full of tables and graphs, this book includes every statistic about an egg that you probably ever could use.

Deeming, D. C. and M. W. J. Ferguson (eds.). 1991. *Egg Incubation: Its Effects On Embryonic Development in Birds and Reptiles*. Cambridge University Press, Cambridge. This is a collection of papers from an international meeting. The comparisons between bird and reptile are fascinating. For example, did you know that although bird eggs require turning during incubation, turning a reptilian egg usually kills the embryo?

Diaz, C., L. Puelles, F. Marín and J. C. Glover. 1998. The relationship between rhombomeres and vestibular neuron populations as assessed in quail-chicken chimeras. *Dev. Biol.* 202: 14–28. This is an excellent example of combining the more modern techniques of immunohistochemistry with the older technique of using chimeras in fate mapping.

Hamburger, V. and H. L. Hamilton. 1951. A series of normal stages in the development of the chick embryo. *J. Morphol.* 88: 49–92. This is the original printing of the Hamburger-Hamilton chick staging series. It contains an exquisite set of photographs and drawings along with the written description of each stage.

North, M. O. 1972. *Commercial Chicken Production Manual*. Avi Publishing Co., Westport, CT. A war-horse of a manual, this will stand by you with ready answers to a host of practical questions about chick eggs and their incubation.

Romanoff, A. L. and A. J. Romanoff. 1949. *The Avian Egg*. John Wiley and Sons, New York. This is a classic. Though some of the biochemical information is now outdated, the wealth of anatomical details is well worth the task of carrying such a heavy book around.

Selleck, M. and C. Stern. 1991. Fate mapping and cell lineage analysis of Hensen's node in the chick embryo. *Development*. 112: 615–626. A superb look at how to do classic fate mapping with modern techniques.

Spratt, N. T., Jr. 1946. Formation of the primitive streak in the explanted chick blastoderm marked with carbon particles. *J. Exp. Zool.* 103: 259–304. This is a classic paper showing the original ways in which fate maps were determined.

Stern, C. D., 1994. The chick. In *Embryos, Color Atlas of Development*. J. B. L. Bard (ed.). Wolfe Publ., London, pp. 167–182. An excellent review of early chick development that includes a discussion of fate mapping techniques.

Tuan, R., 1987. Mechanisms and regulation of calcium transport by the chick embryonic chorioallantoic membrane. *J. Exp. Zool.* [Suppl.] 1: 1–13. This paper demonstrates that calcium needed from bone formation in the chick embryo is resorbed from the eggshell by the chorioallantoic membrane and brought back to the embryo by the allantoic blood vessels.

Suppliers

Prepared microscope slides of chick embryos can be obtained from suppliers such as:

Turtox/Cambosco, Macmillan Science Co., Inc.
8200 South Hoyne Ave.
Chicago, IL 60620
1-800-621-8980

Connecticut Valley Biological Supply Co., Inc.
P.O. Box 326
82 Valley Road
Southampton, MA 01073
1-800-628-7748

Ward's Natural Science Establishment, Inc.
5100 West Henrietta Road
P.O. Box 92912
Rochester, NY 14692-2660
1-800-962-2660
www.wardsci.com

Models of chick development are available from Turtox, listed above.

chapter 10   *33-Hour Chick Embryo*

Whereas the 24-hour chick embryo was roughly equivalent to a 3-week human embryo, a 33-hour chick embryo is roughly equivalent to a 4-week human embryo; what takes hours in a chick embryo takes days in a human embryo. This dramatic difference in developmental rates, in embryos with such similar developmental patterns maintained at roughly the same temperatures, teases the mind with questions. Can you suggest any reasons for the difference? Publish them (I'm serious) or discuss them with your lab colleagues.

You will be using whole mounts and serial cross sections to study the 33-hour chick embryo. Be sure you are able to identify all structures printed in boldface, and record in your laboratory notebook your answers to any questions posed. You will also find it useful to keep track of germ layers as you study the cross sections. Do this by color-coding all the tissues in each cross-sectional diagram in the chapter. Remember the cardinal code of embryological colors: **blue = ectoderm**; **red = mesoderm**; **yellow = endoderm**; **green = neural crest**. The coloring is easy, and the corollaries you can derive from it will serve you well. It teaches you the answers to all those dreaded questions that start out, What is the germ layer origin of...?

Upon looking at the 33-hour whole mount, you will immediately notice that two major advances have taken place since 24 hours of development. The anterior nerve cord is now a brain subdivided into its major regions, and the rudiments of the circulatory system are present. Understand the general outline of these advances to introduce yourself to this stage.

Central nervous system

The neural tube, closed along much of its length, has become specialized anteriorly to form the brain. It has three major subdivisions—the forebrain (**prosencephalon**), midbrain (**mesencephalon**), and hindbrain (**rhombencephalon**)—which are visible as a series of enlargements. The forebrain is further divided into an anterior **telencephalon** and posterior **diencephalon**. The diencephalon is becoming very complex: its walls have evaginated to form **optic vesicles**, which later become the **optic cups**; its floor has evaginated to form the **infundibulum**, which later becomes the posterior part of the **pituitary gland**; its roof, at this point, has no evaginations, but by 48 hours will have evaginated to form the **epiphysis**, later to become the **pineal gland**. This is an astonishing array of derivatives. The hindbrain is also subdivided, forming a **metencephalon** anteriorly and a **myelencephalon**

121

posteriorly. This organization of brain regions is the same as that of the adult brain, but becomes more difficult to discern as differential growth of the regions, most notably of the telencephalon and metencephalon, obscure the embryonic layout.

Circulatory system

The circulatory system has advanced so that there is now a heart in the form of a relatively straight tube with four chambers. It will not start beating until 48 hours of development. At this point, it looks very much like a fish heart (if you have ever mucked through shark oil to study the shark heart, this should sound familiar). Posteriorly, the heart has a **sinus venosus**. This leads forward into an undivided **atrium**, which opens into an undivided ventricle, swung slightly to the right, which in turn leads into the **conus arteriosus** (or **bulbus cordis**). From the conus arteriosus sprout a pair of arteries, the **ventral aortae**.

Think about your own four-chambered heart with its two atria and two ventricles. It started out looking like this embryonic chick heart. What happened to mold it into its adult form? The answer is, very much what happens later in the embryonic chick heart. The atrium and ventricle become subdivided into right and left sides. The sinus venosus becomes reduced to the **sinoatrial node** (or **pacemaker**), and the conus arteriosus becomes subdivided to form the base of the **aorta** and **pulmonary arteries**.

The embryonic blood vessels in the chick at 33 hours are still very rudimentary. There are a pair of **vitelline veins** that will bring nutrient (yolk)-laden blood from the yolk sac back to the embryo. These join the heart at the sinus venosus. A pair of **ventral aortae** leave the heart and connect to the paired **dorsal aortae** by way of the first pair of **aortic arches**. Later additional aortic arches will be added, connecting the ventral and dorsal aortae.

33-Hour Whole Mount

Examine your 33-hour whole mount under a dissecting scope, and compare it to the 24-hour whole mount. Look at the two subdivisions of the **area opaca**: the **area vitellina** and **area vasculosa**. (Again, much of the area vitellina has had to be trimmed away for mounting.) Notice that these regions are considerably larger than at 24 hours. Record how much larger by measuring the widths and lengths of the area vasculosa with your micrometer slide. Have the number of **blood islands** increased as well? At the outer margin of the area vasculosa, there is now a darker band. This is the **sinus terminalis**, a circular blood vessel marking the terminal channel of the vitelline (yolk) circulation. Look at the **area pellucida**, and within the area pellucida the **proamnion** just in front of the head. Recall from the earlier chapter why these areas are translucent, and why the proamnion is particularly clear.

Now examine your slide under the low power of your compound microscope. *Do not use high power.* The slide is too thick and will crack the lens ($$$). Look at the embryonic region to find the structures listed below. (Figure 10.1 will help in your identification.)

Folds and tucks: Morphogenesis proceeds

Head This is huge in relation to the rest of the embryo. Estimate what percentage of the embryo's length is taken up by head as measured by the posterior limit of the brain. The head is elongating faster than the tissues around it and is lifting off the blastoderm anteriorly, forming the head fold in the process.

Lateral body folds In order to create a tubular-shaped body out of a flat plane, lateral body folds tuck down around the embryonic region. You will see these best in cross section.

Primitive streak This is present at the posterior end of the embryo, where gastrulation is still proceeding.

Ectodermal structures

Prosencephalon This is the anterior region of the brain, subdivided into a telencephalon and diencephalon.

Telencephalon This is not distinctly separate from the diencephalon, and it will suffice to label the brain region anterior to the optic vesicles as telencephalon. Later, the telencephalon becomes bilobed and is associated with the olfactory organs and the sense of smell, as well as being the center of intelligence. We more commonly call it the **cerebrum**.

Anterior neuropore This opening at the anterior end of the brain is still present.

Diencephalon This is most easily recognized by its lateral evaginations, the **optic vesicles**. The rest of the lateral walls becomes the thalamus, a region of sensory

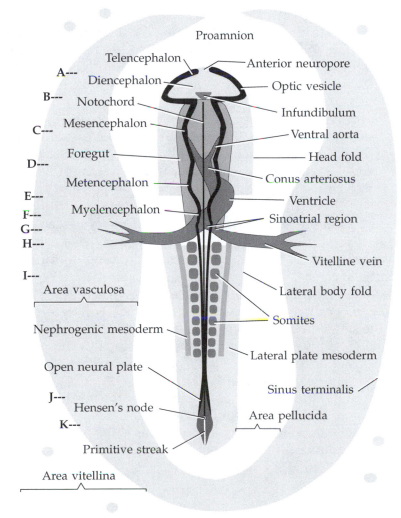

Figure 10.1
Schematic diagram of a whole mount 33-hour chick embryo. Around the edge from anterior to posterior is a series of letters, each marking the level of a cross-sectional diagram shown in Figure 10.2.

integration. The floor becomes the hypothalamus, a region that sends hormones into the posterior pituitary for storage and secretes releasing hormones that control the anterior pituitary. It should come as no surprise, therefore, to find that the posterior pituitary forms from an evagination of the diencephalon floor, the infundibulum. The roof of the diencephalon will form the anterior choroid plexus, a highly vascularized region that secretes cerebral spinal fluid by squeezing out an ultrafiltrate of the blood. An evagination of the diencephalon roof (the epiphysis) forms later, becoming the pineal gland. This is the mysterious "third eye" of some reptiles. (Is this surprising, since the first two eyes are also derived from the diencephalon?) In addition, the pineal gland secretes melatonin and is a seat of control for circadian and seasonal rhythms. In the early 1600s, the famous French philosopher Descartes considered the pineal gland to be the seat of the soul.

Optic vesicles   These are evaginations of the lateral walls of the diencephalon. They grow close to the head epidermis and will induce it to form the lens of the eye. The optic vesicles invaginate to form optic cups, the inner side of which differentiate into the neural retina and the outer layer into the pigmented retina. The optic nerve grows back from the neural retina along the connection of the optic cup to the diencephalon. The optic nerves from opposite sides then cross underneath the diencephalon and enter the mesencephalon. (Think about this: most of your eye is an outpocketing of your embryonic brain.)

Infundibulum   The floor of the diencephalon evaginates, forming the infundibulum. It later differentiates into the posterior pituitary (or neurohypophysis), which stores two hormones from the hypothalamus: antidiuretic hormone (which causes the kidney to resorb water and produce more concentrated urine), and oxytocin (which causes uterine contractions and milk release in mammals).

Mesencephalon   This is the middle, oval-shaped region of the brain that will serve primarily for processing data from the eyes and ears. The dorsal region will become the optic lobes, visual centers associated with the optic nerves.

Rhombencephalon   This region becomes associated with the system of hearing and balance and is divided into an anterior metencephalon and posterior myelencephalon.

Metencephalon   This will form the cerebellum dorsally and the pons ventrally. The cerebellum primarily coordinates stimuli concerning body position and movement; the pons is a bundle of connective pathways shunting information between the cerebrum and cerebellum.

Myelencephalon   Also called the **medulla oblongata**, this region is characterized by a series of enlargements called **neuromeres**, each of which will become associated with a specific set of motor and sensory nerves. The myelencephalon becomes the great linking highway between the brain and spinal cord.

Spinal cord   This is the posterior continuation of the neural tube. Notice how far back on your embryo the closed neural tube extends.

Otic placodes   These are epidermal thickenings on the side of the head in the region of the myelencephalon. "Otic" stands for ear, and these placodes, seen best in cross section, later invaginate to form the inner ear. This is the organ of both hearing (the cochlea) and balance (the semicircular canals).

A category unto themselves: Color them green

Neural crest cells   These are some of the most phenomenal cells in the embryo. The neural crest cells are an evolutionary invention of the vertebrates. First residing on the crest of the neural tube, they later migrate to become a host of structures: pigment cells (except for the pigmented retina), membrane bones in the face and skull, dentine-secreting cells of the teeth (not in birds, of course— why is it, do you think, that birds don't have teeth?), the adrenal medulla, and sympathetic ganglia of the autonomic nervous system are just some of the structures formed. The complete list is much longer. (I've always suspected that if all parts of the vertebrate body disappeared except those derived from neural crest, you'd still have the ghostly image of a body remaining.)

Endodermal structures

Remember that the endoderm forms only epithelial structures. During the formation of the gut, the endoderm contributes only the epithelial tube that forms the inner epithelial lining of the organs of the gut. It is the mesoderm surrounding this endodermal tube that forms the muscle and connective tissue layers of these organs.

Foregut   This is the region of anterior gut that has been closed into a tube. This happens when the head fold and lateral body folds undercut the embryo, closing off the gut and bringing the cardiac mesoderm together beneath (ventral to) the gut. The anterior region of the foregut is called the **pharynx** and forms the esophagus as well as a number of outpocketings, or evaginations. Two evaginations will form ventrally. The anterior one will become the thyroid gland and the posterior one the lungs. A series of lateral evaginations are the pharyngeal pouches that are later transformed into the eustachian tube and cavity of the middle ear, and the epithelium of the tonsils, thymus, parathyroid, and ultimobranchial bodies. The posterior region of the foregut forms the stomach.

Anterior intestinal portal   This is the opening of the foregut into the open midgut. The opening continues to shift posteriorly as closure of the gut continues. It can be seen more readily by turning your slide over and viewing the embryo from the ventral side.

Mesodermal structures

Heart   This lies ventral to the foregut. At this point in development, like the brain, it is huge compared to the rest of the body. (Think where your heart is in relation to your foregut [esophagus]. Is it ventral or dorsal?) Notice that the phrase "having one's heart in one's throat" truly applies to a vertebrate embryo at this stage of development. Obviously, the heart must shift posteriorly to the thoracic region. This will occur primarily through an elongation of the head and neck regions. Look for the large **vitelline veins** coming in from the extraembryonic regions and entering the posterior chamber of the heart, the **sinus venosus**. At this point in development, the sinus venosus cannot be readily distinguished from the next chamber, the **atrium**. Therefore, you can refer to them both as the **sinoatrial region**. The **ventricle** is the next chamber and is easily distinguishable since it bends to the right. Where the heart narrows again anteriorly marks the **conus arteriosus**. If you focus very carefully while viewing the embryo from the ventral side (turn your slide upside down), you may be able to see the paired ventral aortae leading away from the conus arteriosus.

Table 10.1 Staging series of embryonic chick development. (After Hamburger and Hamilton, 1951.)	
Stage and length of incubation (at 38°C)	**Description of embryo**
Stage 1: **Prestreak**	After laying, cleavage is complete and gastrulation has begun, but the primitive streak has not yet formed.
Stage 2: **Initial streak** (6–7 hours)	First formation of the primitive streak.
Stage 3: **Intermediate streak** (12–13 hours)	A broad streak has formed that extends from the posterior margin to the center of the area pellucida.
Stage 4: **Definitive streak** (18–19 hours)	The primitive streak extends about three–quarters the length of the area pellucida, which is now pear-shaped. Hensen's node is present.
Stage 5: **Head process** (19–22 hours)	The notochord is visible, but the head fold has not yet formed.
Stage 6: **Head fold** (23–25 hours)	Anteriorly, the head fold is distinct. No somites have formed yet.
Stage 7: **1 Somite** (23–26 hours)	One pair of somites can be seen, and the neural folds are visible anteriorly.
Stage 8: **4 Somites** (26–29 hours)	Four pairs of somites are present. Neural folds meet at the level of the midbrain, and blood islands are seen in the area vasculosa.
Stage 9: **7 Somites** (29–33 hours)	Seven pairs of somites. The optic vesicles are forming, and the heart is beginning to form.
Stage 10: **10 Somites** (33–38 hours)	10 pairs of somites. Three major subdivisions of the brain are visible. Heart is bent slightly to the right. *eye*
Stage 11: **13 Somites** (40–45 hours)	13 pairs of somites. Five neuromeres of the myelencephalon can be seen. Anterior neuropore is closing.
Stage 12: **16 Somites** (45–49 hours)	16 pairs of somites. Head is turning to the left. Optic placode is invaginating.
Stage 13: **19 Somites** (48–52 hours)	19 pairs of somites. Head is turned almost fully to the side.
Stage 14: **22 Somites** (50–53 hours)	22 pairs of somites. Body is rotated to the side as far back as 7–9 somites. First two pharyngeal pouches and grooves are distinct. Optic vesicles invaginating.

Table 10.1 *(Continued)*

Stage and length of incubation (at 38°C)	Description of embryo
Stage 15: **Optic cup** (50–55 hours)	24–27 pairs of somites. The third pharyngeal pouch and groove have formed. The optic cup is formed.
Stage 16: **Pre-limb bud** (51–56 hours)	26–28 pairs of somites. Tail bud is beginning to form.
Stage 17: **Early limb bud** (52–64 hours)	29–32 pairs of somites. Both wing and leg buds are distinct swellings. Tail is bent ventrally. Allantois not yet present.
Stage 18: **Early allantois** (3 days)	30–36 pairs of somites. Limb buds have enlarged. Tail bud has turned to the right. Allantois is a thick-walled pocket. - *holds nitrogen wastes*
Stage 19: **Somites in tail** (3–3½ days)	37–40 pairs of somites, which extend into the tail. Leg bud is slightly larger than wing bud. Eyes are unpigmented.
Stage 20: **Vesicular allantois** (3–3½ days)	40–43 pairs of somites. Leg bud slightly asymmetrical. Allantois vesicular, about the size of the midbrain.
Stage 21: **Early eye pigmentation** (3½ days)	43–44 pairs of somites. Both wing and leg buds slightly asymmetrical. Allantois enlarged. Eye pigmentation faint.
Stage 22: **Distinct eye pigmentation** (3½–4 days)	Somites extend to tip of tail. Eye pigmentation is distinct.
Stage 23: **Limb buds as wide as long** (4 days)	Both wing and leg buds are as wide as they are long.
Stage 24: **Limb buds longerthan wide** (4½ days)	Both wing and leg buds are distinctly longer than they are wide. Toe plate is distinct, but toes are not yet demarcated.
Stage 25: **Early elbow and knee** (4½–5 days)	The elbow and knee joints are distinct. The digital plate in the wing is distinct but with no demarcation of digits. Third toe is demarcated by a faint groove.
Stage 26: **Three toes** (5 days)	Limbs are considerably longer. The demarcation of three toes is distinct. There is a faint groove between the second and third digit of the wing.
Stage 27: **Early beak** (5–5½ days)	The three digits of the wing are demarcated by grooves. The beak is just recognizable.
Stage 28: **Beak distinct** (5½–6 days)	The second digit of the wing and the third toe are longer than the rest. The three wing digits and four toes are distinct. (Chickens only have three "fingers" and four toes.) The beak is a distinct outgrowth.

Notochord   This is visible as a dark line in the midline ventral to the neural tube. Determine how far anteriorly and posteriorly it extends. Pause to reflect on its importance. Remember, chordates wouldn't exist without the notochord. The notochordal mesoderm has been called the "primary organizer" because of its many inducing functions that initiate an axial organization of the organism. It also provides axial support for the embryo until the vertebrae engulf it, taking over this function. In the adult, notochordal tissue between the vertebrae expands to form the gelatinous center (nucleus pulposus) of the intervertebral disc. When a person suffers from a "slipped disc" or herniated disc, stress on the back has caused the nucleus pulposus to bulge outward. The bulging disc puts pressure on the spinal cord and causes back pain. Often, rest and proper back exercises can remedy this condition without the need of surgery. There is also a core of notochord left in the centrum of each vertebra. Sometimes this remnant of notochord gives rise to tumors called chordomas. These tumors cause extreme back pain by narrowing the spinal canal and pinching the spinal cord and must be removed surgically.

Somites   These blocks of mesoderm are found on either side of the neural tube. Each somite becomes subdivided into a **sclerotome** that forms the vertebrae and ribs, a **myotome** that forms skeletal muscles for the back and limbs, and a **dermatome** that gives rise to dermis. Count the number of pairs of somites your embryo has and use this to stage your embryo using the Hamburger-Hamilton staging series (Table 10.1). Record the stage.

Nephrogenic (intermediate) mesoderm   This is a streak of mesoderm lying lateral to the somites. It will form the kidneys, the urogenital ducts, and the gonads.

Lateral plate mesoderm   Lateral to the nephrogenic mesoderm is the lateral plate mesoderm, which delaminates to form two layers: the **splanchnic** and **somatic** layers. The splitting creates a cavity, the body coelom. Somatic lateral plate mesoderm becomes associated with the epidermal ectoderm, and together they form the somatopleure. This later will form the body wall and, in the extraembryonic regions, the **amnion** and **chorion** (the extraembryonic membranes that surround and protect the embryo). The splanchnic lateral plate mesoderm associates with the endoderm, which together form the splanchnopleure. This forms the gut wall and, in the extraembryonic regions, the vascularized **yolk sac** and **allantois**. Because of its extensive vascularization, the yolk sac can transport yolk nutrients back to the embryo, and the allantois can receive nitrogenous waste from the embryo for storage as uric acid.

33-Hour Serial Sections

The ability to read serial sections and to construct a three-dimensional mental image of the whole is a skill worth acquiring. Be patient with yourself. Look at your slides; there will probably be two of them. The embryo has been sliced into (usually) 10-µm-thick sections, and the sections have been placed in order, anterior to posterior. Not a single section is lost. There will be a number of rows of sections on each slide. By convention, these were laid down starting at the upper left-hand corner of the first slide. In scanning the slide, you will read it just as you would a book, from left to right. Now put the slide on your compound microscope and move the controls of the stage so that the slide moves from left to right along the top row. Notice that when you look through the microscope, your movements look reversed, and the slide appears to be moving from right to left. To read the next row, you must move the slide down and all the way back to the left. All this magnified motion whizzing past your eyes can make you feel sick in short order. Some hints from a master of queasiness: never look through the microscope as you are changing rows, and always move slowly from one section to the next, relaxing your eyes as you do to avoid trying to focus on the blurred image as it passes.

Look at the diagrams of cross sections (Figures 10.2A–K). The letter of each cross section A–K matches a letter on the whole-mount diagram (Figure 10.1), showing the level from which each cross section was taken. Your own sections will probably look slightly different from the diagrams, since the angle of cutting can vary. As you read through your serial sections, it is extremely important that you constantly refer back to the whole mount to know where you are in relation to the rest of the embryo. Remember to color-code the cross-sectional diagrams. The letters below correspond to the letters of the cross-sectional diagrams in Figure 10.2.

Section A Start reading your sections from the anterior end. The first few sections through the **prosencephalon** will show a narrow ventral cleft through the neural tube. This is the **anterior neuropore**. Identify the two-layered **proamnion** just below the head. What are the two germ layers represented here? There is a space between the head and the proamnion called the **subcephalic** ("below the head") **pocket**. Remember why this space exists: the head fold has undercut the embryo, lifting the head up from the level of the blastoderm. Lateral to the proamnion, the **somatopleure** and **splanchnopleure** can be seen, with the large **coelomic space** between them. Notice that there are **capillaries** in the mesoderm of the splanchopleure. This layer becomes highly vascularized. The capillaries you see are part of the vitelline circulation.

Section B Sections through the **diencephalon** will show the lateral **optic vesicles** and ventral **infundibulum**. Look carefully at the **epidermal ectoderm** that is in contact with the optic vesicles to see if there are any areas of thickening. If there are, you have found the **lens placodes**. These later will invaginate to form the lens vesicles. The loose tissue surrounding the brain is **head mesenchyme**, derived from mesoderm and neural crest cells, which will later form such structures as the skull, head musculature, blood vessels, and connective tissue.

Section C Further posteriorly you will see the rounded cross section of the mesencephalon. At this point, the **foregut**, looking like a smile, should be obvious below the **notochord**. The foregut in this region comes in contact with the epidermal ectoderm to form the **oral membrane**. Fortunately, this membrane is removed by programmed cell death; the opening produced is the mouth opening. When cell death fails to occur, an anomaly results called astomia ("no mouth"). Above the foregut you should see the paired dorsal aortae, and below the foregut the paired ventral aortae. A pair of **first aortic arches** connects the ventral aortae with the dorsal aortae. Which way will the blood flow through these vessels? Show the direction by drawing arrows on the diagram.

Section D Trace the ventral aortae posteriorly and see that they fuse to form the **conus arteriosus**. The head is now attached to the blastoderm, but the indentations of the lateral body folds can still be seen. These eventually will undercut the embryo and lift it off the blastoderm. Notice the narrowed **rhombencephalon**. The anterior portion of this is the **metencephalon**.

Section E If you are very observant as you move posteriorly, you will be able to make

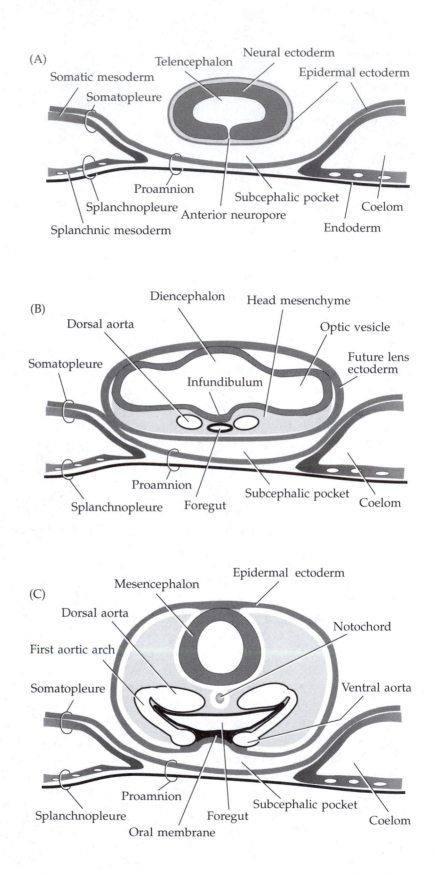

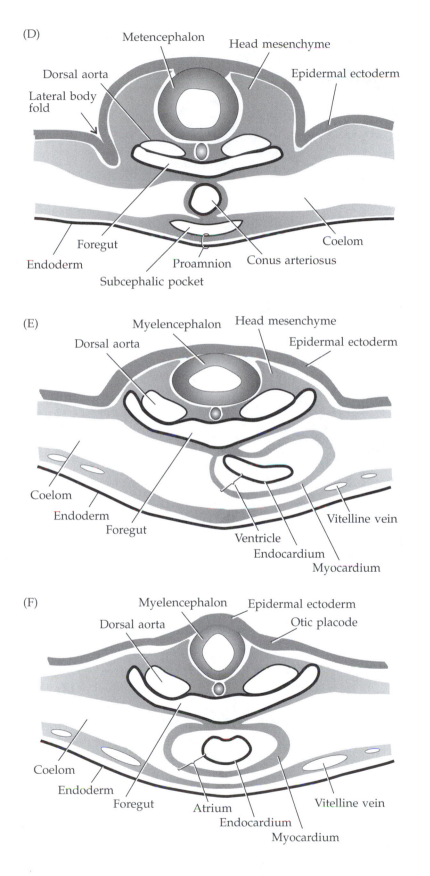

(D) Metencephalon Head mesenchyme Dorsal aorta Epidermal ectoderm Lateral body fold Foregut Coelom Endoderm Proamnion Conus arteriosus Subcephalic pocket

(E) Myelencephalon Head mesenchyme Dorsal aorta Epidermal ectoderm Coelom Endoderm Foregut Ventricle Endocardium Myocardium Vitelline vein

(F) Myelencephalon Epidermal ectoderm Dorsal aorta Otic placode Coelom Endoderm Foregut Atrium Endocardium Myocardium Vitelline vein

Figure 10.2
Cross-sectional diagrams of the 33-hour chick embryo. Diagrams are arranged from anterior to posterior. The letter beside each diagram corresponds to the level shown on Figure 10.1 from which the section is taken.

(G)

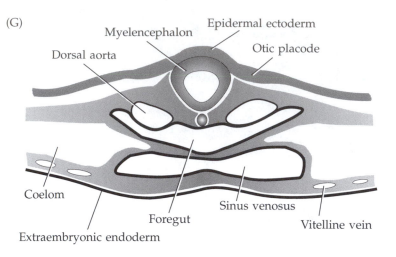

(H)

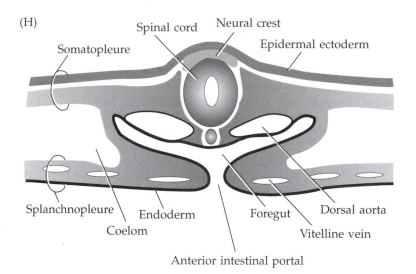

out the intermittent expansions of the **neuromeres** as you go through the **myelen-cephalon**. At the level of the myelencephalon, you also should see that the heart appears larger and swings out to the right. This region of the heart is the **ventricle**.

Section F   As you move posteriorly through the heart, it swings back to the midline. You are now in the region of the **atrium**. Look carefully at the dorsal epidermal ectoderm in this region. On either side of the myelencephalon, you should see epidermal thickenings. These are the **otic placodes**, which will later invaginate to form the inner ear.

Section G   Further posteriorly, where the heart suddenly flattens out, you are in the region of the **sinus venosus**. The **vitelline veins**, feeding into the sinus venosus, can be seen in cross section throughout the mesoderm of the **splanch-nopleure** in this region.

Section H   Just posterior to the region of the sinus venosus, the foregut opens at the **anterior intestinal portal** to become the open **midgut**. Look at the neural

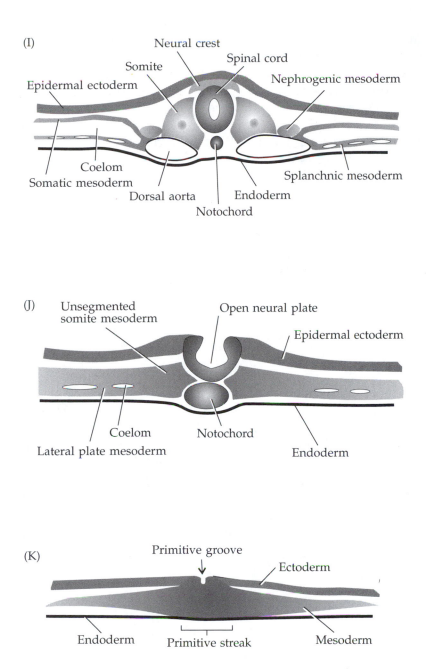

(I)

Neural crest
Spinal cord
Somite
Nephrogenic mesoderm
Epidermal ectoderm
Coelom
Somatic mesoderm
Splanchnic mesoderm
Dorsal aorta
Endoderm
Notochord

(J)

Unsegmented
somite mesoderm
Open neural plate
Epidermal ectoderm
Coelom
Notochord
Lateral plate mesoderm
Endoderm

(K)

Primitive groove
Ectoderm
Endoderm
Primitive streak
Mesoderm

tube and for the tightly packed cells dorsally, on either side of the neural tube. These are the **neural crest cells**. Prior to their migration, they can be seen all along the dorsal side of the neural tube.

Section I   Once you are in the region of the open midgut, you are past the brain and in the area of the **spinal cord**. Here you should see clearly the subdivisions of the mesoderm: the **somites**; more laterally, the **nephrogenic mesoderm**; and finally, the **lateral plate mesoderm**, which is split into **somatic** and **splanchnic** layers. Go back and forth through several somites to see that somites are indeed blocks of tissue and not a continuous rod. Count the number of somites as you travel posteriorly through them and use this to stage your embryo. How close is it to the age of your whole-mount embryo? Record the stage in your laboratory notebook.

Section J   As you proceed posteriorly, the neural tube will become an **open neural**

plate, and the **somite mesoderm** will be unsegmented. Notice that the **lateral plate mesoderm** is just beginning its delamination to form the coelom.

Section K   Still further posteriorly, you will reach the region of the **primitive streak**, where ingression is still occurring. Because differentiation in the chick embryo proceeds in an anterior-to-posterior wave, you will find that traveling posteriorly through the sections gives the impression of traveling backward in time. If you want to get the impression of traveling forward through time, study your sections from posterior to anterior.

Experiment with Evolution

Make a diagram of a cross section through the chick embryo in the region of the somites. Now cut it out, and keeping it flat on a table, bend the two ends around until they meet ventrally. Compare this to a diagram of a cross section through an amphibian neurula. What does this tell you about patterns of gastrulation and the impediments of yolk?

Create a Puppet

Visualizing the tucks and folds that create the chick embryo out of the three-tiered gastrula sandwich is difficult without models. You can greatly enhance your understanding by making a puppet out of three layers of cloth, an outer blue, middle red, and lower yellow (of course). Put your arm under the layers of cloth. Lift your hand slightly, and tuck the cloth in around your hand. You have just created a **head fold**. Look at the underside of your puppet, and see how the folding has created an anterior tube lined in yellow (the **foregut**). Now tuck the cloth in along the sides of your arm to create **lateral body folds**. Wherever you tuck it in completely so that right side meets left side, you have closed off more regions of gut and lifted the embryo up above the level of the rest of the cloth (the **extraembryonic regions**).

Accompanying Materials

Tyler, M. S. and R. N. Kozlowski 2000. *Vade Mecum: An Interactive Guide to Developmental Biology.* Sinauer Associates, Sunderland, MA. "Chick Mid." This chapter of the CD includes whole mounts, color-coded to show germ layers, and labeled, with definitions of anatomical terms. Also a complete set of serial cross sections through a 33-hour chick embryo includes labels and color-coding to show germ layers.

Gilbert, S. F. 2000. *Developmental Biology*, 6th Ed. Sinauer Associates, Sunderland, MA. Chapters 12–17. These chapters have excellent diagrams and descriptions of neurulation, and somite, heart and gut formation, as well as limb and gonad development. In chapter 15 there is a diagram of an amphibian neurula that can be used for the "Experiment With Evolution" section above.

Fink, R. (ed.). 1991. *A Dozen Eggs: Time-Lapse Microscopy of Normal Development.* Sinauer Associates, Sunderland, MA. Sequence 11. This shows head fold, brain, somite, and heart formation, along with closure of the foregut and regression of Hensen's node.

Selected Bibliography

Hall, B. K. and S. Hörstadius. 1988. *The Neural Crest.* Oxford University Press, London. This is a detailed, thought-provoking volume on the evolutionary origins, migrations, and differentiation of neural crest cells. It includes a facsimile reprint of the classic book by S. Hörstadius from 1950.

Hamburger, V., and H. L. Hamilton, 1951. A series of normal stages in the development of the chick embryo. J. Morphol. 88: 49–92. This is the original publication of the Hamburger-Hamilton chick staging series.

Hamilton, H. L. 1952. *Lillie's Development of the Chick. An Introduction to Embryology.* Henry Holt and Co., New York. This revised version of F. R. Lillie's original, much longer, book is a classic. It contains a full reprint of the Hamburger-Hamilton chick staging series.

Kil, S. H., C. E. Krull, G. Cann, D. Clegg and M. Bronner-Fraser. 1998. The α_4 subunit of integrin is important for neural crest cell migration. *Dev. Biol* 202: 29–42. This is an excellent example of how modern techniques in immunohistochemistry are used to track neural crest cells and determine the components important to their migration.

LeLièvre, C. and N. M. LeDouarin. 1975. Mesenchymal derivates of the neural crest: Analysis of chimaeric quail and chick embryos. *J. Embryol. Exp. Morph.* 34: 125–154. These two authors have contributed impressively to our knowledge of neural crest migration and differentiation. This study shows how cell migration can be mapped using embryonic cells from the quail grafted into a chick embryo. The quail cells can be distinguished since their nuclei stain more darkly than those of the chick.

Lehman, H. E. 1987. *Chordate Development*, 3rd Ed. Hunter Textbooks, Winston-Salem, NC. This tome on vertebrate development is filled with interesting detail. In addition to describing developmental anatomy, it discusses development from an evolutionary perspective.

Mathews, W. W. and G. C. Schoenwolf. 1998. *Atlas of Descriptive Embryology*, 5th Ed. Prentice-Hall, Inc., Upper Saddle River, NJ. This traditional manual is one of the best for its photographs of embryos and includes a glossary of terms.

Noden, D. M. 1980. The migration and cytodifferentiation of cranial neural crest cells. In *Current Research Trends in Prenatal Craniofacial Development*, R. M. Pratt and R. L. Christiansen (eds.). Elsevier/North-Holland, New York, pp. 3–26. This author has done painstaking, precise fate-mapping, revolutionizing our understanding of the skull. That fate mapping is still so important to embryology is evidenced by Noden's work and is clearly encapsulated in this review paper.

Romanoff, A. L. 1960. *The Avian Embryo.* Macmillan, New York. This is the biggest book you'll ever want to read on the bird embryo. The diagrams and charts are extensive. Simply keep in mind that some ideas concerning germ layer derivatives (those of the neural crest, for example) have changed in recent years.

Suppliers

See previous chapter.

11 The Living Embryo and Making of Whole Mounts

72- and 96-Hour Chick Embryo

So far you have studied mounted specimens of 24- and 33-hour chick embryos. You now will be observing living chick embryos at 72 and 96 hours of incubation (comparable to the fifth week of development in humans) and learning how to make whole mounts of your own. Your whole mounts can be every bit as good as, and in some cases better than, those bought from laboratory supply houses. You will notice that the procedures for making whole mounts cannot be completed in a single day. Break-points, where you can let your specimens sit until the following laboratory period, are clearly indicated. It will take three laboratory sessions to complete your slides. In the second and third sessions, however, the manipulations are brief and will not interfere with your other laboratory assignment for that week. Just don't forget to do them.

In proceeding through the exercises, it is advisable to complete observations on one embryo and then fix it before going on to your second embryo. Leave the second embryo in the incubator until you are ready to use it.

Incubation of Eggs

Most likely, eggs already have been incubated for you. However, as a scientist, you must be able to repeat all aspects of a study, so here are the basics of egg incubation.

Eggs are normally incubated in a forced-draft (it has a fan), humidified (57% relative humidity) incubator at 37.5–38°C. These incubators come in all levels of sophistication. The old-fashioned, cedar-box incubators, still used by many small chicken farmers, consist of a thermostated heating coil with a small fan that circulates air within the box. Humidity is controlled very simply by placing pans of water in the incubator. These incubators must be checked daily to add water and manually turn the eggs. Turning prevents the embryo from becoming stuck to the shell. Modern mechanized incubators don't require daily checking, since they include an automatic water supply and turning device. Very simple models, however, can also work well. I have even seen boxes rigged with lightbulbs on a rheostat that work perfectly adequately, especially if the eggs do not have to be maintained until hatching.

However simple or complex, the incubator must have good temperature control. Though 37.5°C (99.5°F) is ideal, eggs will tolerate temperatures between 35°C and 40.5°C (95–105°F). Within this range, the cooler temperatures significantly slow the rate of development, and the warmer temperatures accelerate it. A humid atmosphere is also essential in order to avoid dehydration of the egg. Though 57% relative humidity is considered ideal, eggs will tolerate a range of 50–60%. Relative humidity can be determined by reading the difference between a wet-bulb and a dry-bulb thermometer. The dry-bulb is a normal thermometer. The wet-bulb is simply a thermometer whose bulb is covered by a wick kept moist with water. This records a cooler temperature due to evaporation from the wick. The less humid the air, the greater the cooling. When the incubator is 37.5°C (99.5°F) and the wet-bulb thermometer reads 30°C (86°F), the relative humidity is approximately 57%. Normally, just keeping a large pan full of water in the incubator is sufficient to give the appropriate humidity. Check this pan daily. It should never go dry.

Seldom is it convenient to start incubation on the day that an egg is laid. Luckily, eggs can be stored prior to incubation, though hatchability diminishes as the storage time increases. (By 22 days of storage at 10.5°C, hatchability is down to 26%, and by 25 days, it is 0%.) If the eggs are to be stored for less than 14 days, the temperature of the holding room should be 18.3°C (65°F), and the humidity should be as high as 80%. If they are to be stored for longer than 14 days, the holding room should be 10.5°C (51°F). When the eggs are removed from the holding room, they should warm to room temperature before being placed in the incubator.

Eggs should be incubated with their blunt side uppermost. This can be done most easily by leaving the eggs in the cardboard flats that they come in. The eggshell should be labeled in pencil to indicate the day the egg is put into the incubator. For proper development, eggs should be turned several times a day—as many as six. It is best to turn them to a 45° angle from their vertical position. If incubating for only 3 or 4 days, as in this study, turning is unnecessary.

During incubation, eggs can be candled to determine whether they are developing or not. A candler is no more than a light in a box shining through a hole the diameter of an egg. You can make a candler by putting a light in a wooden box that has a hole the size of an egg cut in one end. Candling is best done in a dark room. Move the egg around, holding it against the lighted hole of the candler. In the living embryo, vascularized regions will be particularly obvious. A dead embryo will look dull and dense.

If the eggs are being kept to hatching, the temperature of the incubator ideally should be reduced a degree or two during the last 2 days of incubation (it takes 21 days to make a chick). The chicks will be breaking through the blunt end of the shell. After the chick has "pipped" the shell (made the first, small hole), it will rest for hours or even days before finishing the job, which may take as little as an hour.

Preparing Whole Mounts

The eggs you will be using for whole mounts should not be turned while incubating and should have been positioned with their blunt end up. Because the yolk orients to gravity with the embryo uppermost, the embryo should now be sitting just below the air space at the blunt end of the egg. This makes it easily accessible for observation and dissection.

Get one of your eggs that has been incubated for 3 or 4 days from the incubator and place it in a "professional egg-dissection holder" (i.e., the bottom half of an egg carton cut to contain three or four depressions for eggs). Make sure that the blunt end of your egg is still facing upward. (You may candle it first to determine exactly where your embryo lies.) You will be opening the egg simply by attacking it as you would a really soft boiled egg at breakfast. With the handle of your forceps, tap the blunt end of the egg five or six times, cracking a nickel-shaped region of shell into a number of small pieces. Remove the broken bits of shell with the forceps. The outer shell membrane will be removed with the shell. Now with fine forceps, very carefully remove the tough white inner shell membrane overlying the embryo.

   The embryo is now exposed. Observe the embryo under the dissecting scope. The heart should be beating. Notice the jerkiness of blood movement in the arteries and the smooth flow in the veins. Use Figures 11.1 and 11.2 to identify embryonic structures and extraembryonic regions. You can make a number of observations before proceeding further. Ask as many questions as you can think of. Record both your questions and answers in your notebook. For example, what is the heart rate? Use a watch with a second hand to record rates. Is there both an atrial and ventricular beat? Do you see skeletal muscular movements? Does the embryo respond to touch? Does the amnion or chorion respond to touch? Does warming the egg under a lamp increase or decrease muscle or heart movement? Does cold Ringer's solution? Record any rate changes. Does alcohol affect the

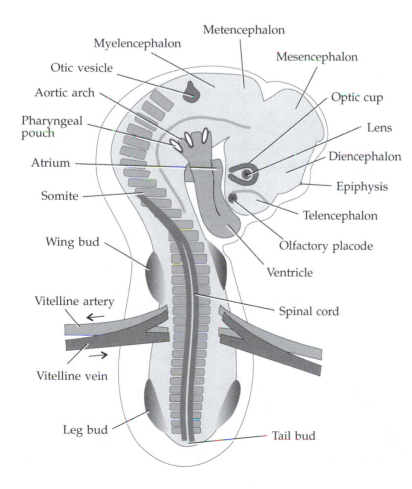

Figure 11.1
Schematic diagram of a 72-hour chick embryo.

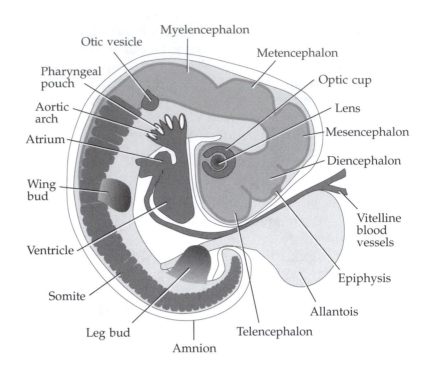

Figure 11.2
Schematic diagram of a
96-hour chick embryo.

heartbeat? You could test ranges that are found in the blood of a person considered legally drunk (0.1% or greater; at levels of 0.05–0.09%, a person is considered "under the influence"). Remember that alcohol is a fixative—don't use higher percentages that might kill the embryo. Does caffeine or theophylline alter the heartbeat or muscle movements? You can test this using dilutions of the pure chemicals or with drinking strengths of coffee (containing caffeine) and tea (containing caffeine and theophylline). A good cup of coffee contains approximately 0.5–1 mg of caffeine per ml, with instant having the least and drip-prepared having the most. A cup of tea has about 0.3 mg of caffeine and 0.04 mg of theophylline per ml. Remember what these chemicals do. They increase cyclic adenosine monophosphate (cAMP) levels in cells by inhibiting the enzyme phosphodiesterase that breaks down cAMP. Increasing or decreasing cAMP levels can have profound effects on a cell. Cyclic AMP levels increase in a cell that is no longer dividing, for example, and such increases precede programmed cell death. Low levels of cAMP occur in dividing cells and are maintained at these low levels in cancer cells. What do you imagine the effect is on an embryo (in which cell division is so essential) of raising cAMP levels over long periods of time? What is your advice to pregnant women on drinking alcohol, coffee, or tea? (*N.B.:* None of the substances listed should be injested by a pregnant woman. In particular, no level of alcohol is considered safe for the embryo or fetus.)

Think of other interventions you might do to determine more about the embryo. Record all your observations in your laboratory notebook. Organize your data into graphs wherever possible. What conclusions do you draw from your data?

You are now ready to remove the embryo from the yolk. This often requires three hands, especially on the first try. You will need to cut the embryo free from its extraembryonic membranes while holding on to it, so that it doesn't slip into the murky depths of the yolk once it is freed. If the embryo is large enough to have an identifiable neck region, you can slip your forceps underneath the neck. With your other hand and a pair of very sharp scissors, cut the extraembryonic membranes around the embryo. Now slip your embryo spoon under the embryo. This is where a third hand standing by often saves the day. Do not let go of the embryo with the forceps, even though the embryo seems safe on the spoon. Lift the embryo carefully from the yolk. Often as you lift the embryo away, you will see that it isn't completely free from the membranes. Have scissors ready to cut any remaining strands. Transfer the embryo to a petri dish containing Ringer's solution (a balanced salt solution).

Chick embryo Ringer's solution

NaCl	7.20 gm	*Make up to 1 liter in glass*
KCl	0.37 gm	*using distilled water.*
CaCl$_2$	0.17 gm	

If your embryo is too young to have an identifiable neck, an easy way to remove it is first, to use your forceps to grab on to the extraembryonic blood vessels that feed into the heart–and don't let go. Then, use sharp scissors to cut around the embryo, cutting the extraembryonic membranes so that the embryo is freed from the yolk. Without moving the embryo, put the scissors down and pick up your embryo spoon (or use a partner to provide a third hand). Slip the embryo spoon under the embryo as you gently pull the embryo onto the spoon using the forceps (that are still holding onto those blood vessels). Now pull the whole thing away (still holding on with the forceps), having scissors ready to snip any strands of membrane that are still attached. Place the embryo into a petri dish containing Ringer's solution.

Make observations on the embryo removed from its yolk. You should be able to distinguish more features. Stage your embryo using the Hamburger-Hamilton staging series (see Table 10.1). In your laboratory notebook, record the stage and the number of somites your embryo has. Observe the heartbeat again. Has it changed? If you warm the embryo under a lamp, does this change the heartbeat? This is your last chance to make as many observations as you'd like on the living embryo. Be creative. When you have completed your observations, go on to the fixation step, then repeat the above steps on your second embryo.

Fixation

Fixation is the first step in any procedure in which tissue is to be preserved for histological study. Fixatives kill. They kill the tissue as well as any bacteria that are present that otherwise would cause the tissue to rot. They also coagulate proteins, making them insoluble, and strengthen protein linkages. All fixatives distort the tissue to a certain extent, but in general, protein and cell structure are preserved. You will normally choose a fixative containing several ingredients that balance out the ill-effects of each other. For example, alcohol shrinks tissue and causes excessive hardening. This can be countered by adding acetic acid, which swells tissue and prevents overhardening.

Carnoy's fixative

glacial acetic acid	100 ml
100% ethanol	300 ml

Glacial means 100% acetic acid; vinegar is 5% acetic acid. Hard liquor such as whiskey is about 40%–50% ethanol. Rubbing alcohol is 70% isopropyl alcohol— don't drink it, it's toxic. If you were out in the boondocks with no scientific supplies and found the perfect specimen you needed to preserve, what makeshift fixative would you devise?

Carnoy's fixative, a mixture of alcohol and acetic acid, is a widely used fixative. It is not ideal; the addition of formaldehyde, for example, would give better preservation of cytological detail. But Carnoy's fixative will not leave any toxic residues in the lab. Formaldehyde and most other fixing agents leave highly toxic residues that are virtually impossible to remove from instruments and glassware. In making whole mounts, where cytological detail is not critical, Carnoy's becomes ideal for our use because there is no risk of contaminating our lab for future live material. (You may be using one of your embryos for the histological sectioning procedures discussed in the next chapter. If so, you might be fixing in something other than Carnoy's. Check with your instructor before proceeding.)

With a pipette, drop fixative on top of the embryo until all regions have been covered. This will allow any extraembryonic membranes to fix flat before flooding the specimen with fixative. After 10 minutes, flood the embryo with fixative. Wait another 10 minutes, then transfer the embryo to a screw-cap vial. For fixation, and for all steps that follow that require a fluid, the rule of thumb is that the amount of fluid used should be 10 times the volume of the tissue. Label your vial with your name and the stage of your embryo. Always use pencil for labeling—it doesn't come off in the reagents.

Your embryos should fix for a minimum of 2 hours. They can be left in the fixative a maximum of 2–3 days. Following fixation, your embryos will be washed and stained, then dehydrated and cleared before mounting. All of these operations can be done in the vial. To make a transfer of solution, you can either pour off the old solution (*always over a finger bowl—never over the sink*; guess why), or you can pipette off the old solution. The new solution is added directly to the vial. This avoids unnecessary handling that can damage the embryo.

Washing

You will be washing out the fixative using 70% ethanol. Washing is often done in water. If you have used Carnoy's as a fixative, washing in 70% alcohol is both adequate and fast. You also can use it as a holding solution. The embryos can remain in 70% alcohol indefinitely without harm.

Remove the fixative and add 70% ethanol to your vials. Your embryos should be in this for a minimum of 8 hours. Keep the embryos in 70% ethanol until next week's lab. You should change the solution at least once during the week to ensure that all fixative is washed out of the tissue.

---

* Some Carnoy's recipes also include chloroform, which is extremely hazardous. Note that under current OSHA guidelines, solutions containing concentrations of alcohol above 24% must be discarded as hazardous waste.

Under current OSHA guidelines, any waste alcohol solutions containing over 24% alcohol must be handled as hazardous waste and may not be poured down the sink.

Staining (To Be Done the Week Following Fixation)

You will be using a whole-mount staining procedure that is exceedingly stable and will not overstain your embryos. Before staining, pour off the 70% alcohol and add distilled water to your vials. Let the embryos soak for at least 10 minutes before transferring them to the stain. Then pour off the water and add Mayer's carmalum stain. This is a general nuclear stain that is excellent for whole mounts. Stain for about 48 hours. The timing is not especially critical. The embryos should be a fairly deep red. Once the desired level of staining is reached, pour off the stain and add 70% ethanol to the vials. You will be storing the embryos in 70% ethanol until next week's lab.

Mayer's carmalum

Stock solution:

carmine	1 gm
ammonium alum (aluminum ammonium sulfate)	10 gm
distilled water	200 ml

Working solution:

carmalum stock solution	5 ml	*Mix. When dissolved, filter. Add 1 ml*
glacial acetic acid	0.4 ml	*formalin to inhibit mold growth.*
distilled water	100 ml	*Will keep indefinitely.*

Dehydration and Clearing (To Be Done the Week Following Staining)

It is necessary to dehydrate the embryos because the mounting medium used here will not mix with water. Rather than removing the water from your specimens in a single step, which could damage the tissue, you will use a graded series of alcohols to slowly replace the water with alcohol. This avoids severe convection currents that are set up as one fluid replaces the other.

Following dehydration, the embryos must be cleared in a clearing agent. You could use xylene or toluene. Toluene is gentler on the tissue, but is more expensive, and in making whole mounts the gentleness of the clearing agent is not critical. Clearing will make the embryos transparent. Clearing is also necessary because the mounting medium will not mix either with water or with alcohol. Xylene or toluene, which are immiscible with water, will mix with both 100% alcohol and the mounting medium. These clearing agents therefore provide a bridge for us between the dehydration steps and mounting.

Dehydrate and clear your specimens according to the following schedule. For each transfer of fluid, pour off the old fluid and add the new directly to the vials. Used alcohol at concentrations above 24% must be discarded as hazardous waste according to OSHA guidelines. Xylene and toluene are both very toxic and also must be disposed of as hazardous waste. These agents dissolve plastic and will dissolve plastic pipes if thrown down the sink; they will also dissolve plastic petri dishes or vials, so make sure the vials you are using are glass. Because of their toxicity (they rot your liver), **the clearing step must be done under a fume hood,** or do it outside.

If, on your first transfer into xylene, you notice any white clouding of the fluid, this means that there is still water in the tissues. You must immediately go back to another 15-minute change of 100% ethanol. If you don't, there will be water droplets in your preparation that will make your slide look like salad dressing.

Dehydration and clearing schedule

70% ethanol	10 minutes
95% ethanol	two changes, 10 minutes each
100% ethanol	two changes, 15 minutes each
xylene	two changes, 10 minutes each

Mounting

The mounting medium you use should be a neutral mounting medium, such as Histoclad. Cheaper mounting media are acidic. They cause the stain to fade and oxidize very quickly, making a cloudy ring around the edge of the slide that gradually works its way to the middle. Any slide that needs to be permanent is worth the extra cost of a neutral mounting medium.

The hardest part of mounting is devising a method to raise the coverslip enough so that it isn't resting directly on the specimen. If your specimen is thin, you can use chips of coverslip as a spacer between the slide and coverslip. If your specimen is thick, pieces of thicker glass such as a slide could be used. I have had wonderful success using slices of thin polypropylene tubing cut into doughnuts that are the thickness of the specimen. I have even used metal washers, though plastic washers would be better. Devise a method that will work for you.

Once you start the mounting procedure, work with speed. *Do not let your specimen dry out.* It will do so very quickly. First, with a diamond marking pen, scratch your initials and the stage of your embryo onto one end of a slide. Then lift your embryo from its vial of xylene and place it on the slide. Position the spacer around the embryo. Put a generous amount of mounting medium over the specimen and carefully lower a square coverslip into place, letting one side hit the mountant first and then slowly lowering the rest of the coverslip. If you see that there isn't enough mounting medium, more can be introduced from the side by placing extra mountant alongside the coverslip and letting it wick underneath. If you have too much mounting medium and it is oozing out, *don't touch it.* After it dries for a week, excess mountant can be removed with a razor blade (**carefully**).

Allow your slides to dry flat in a dust-free environment (such as a small cardboard box) at room temperature for at least a week before observing them under the microscope. When you do observe them, restage the embryos to see if your staging of the living embryos was correct. It is far easier to stage an embryo in a whole mount than one that is unstained and uncleared. You now should have some very beautiful, permanent slides.

Accompanying Materials

Tyler, M. S. and R. N. Kozlowski 2000. *Vade Mecum: An Interactive Guide to Developmental Biology*. Sinauer Associates, Sunderland, MA. "Chick Late." This chapter of the CD shows movies of the technique for removing the embryo from the yolk, as well as labeled whole mounts of 72- and 96-hour chick embryos, and a series of movies about the living embryo.

Selected Bibliograpy

Hamburger, V. and H. L. Hamilton. 1951. A series of normal stages in the development of the chick embryo. *J. Morphol.* 88: 49–92. This is the original publication of the Hamburger-Hamilton chick staging series.

Humason, G. L. 1972. *Animal Tissue Techniques*, 3rd Ed. W. H. Freeman, San Francisco. This is one of the best sources on histological techniques—it is out of print, but has been reissued by John Hopkin's University Press (see reference below).

Presnell, Janice K., Gretchen L. Humason and Martin Schreibman. 1997. *Humason's Animal Tissue Techniques*, 5th Ed. Johns Hopkins Univ. Press, Baltimore. The reissuing of Humason's classic.

Romanoff, A. L. 1960. *The Avian Embryo*. Macmillan, New York. This book has an extensive chapter on the embryonic heart, including experimental results on changes in heart rate under conditions of altered temperature and chemical environment.

Suppliers

Fertile chick eggs are most easily obtained from local hatcheries. If this is not possible, eggs can be shipped from supply companies such as those listed below. These companies also have inexpensive egg incubators. More expensive, cedar-box incubators can be bought from farm suppliers.

Connecticut Valley Biological Supply Co., Inc.
P.O. Box 326
82 Valley Road
Southampton, MA 01073
1-800-628-7748

Nasco
901 Janesville Ave.
Fort Atkinson, WI 53538- 0901
1-800-558-9595
www.nascofa.com

Ward's Natural Science Establishment, Inc.
5100 West Henrietta Road
P.O. Box 92912
Rochester, NY 14692-2660
1-800-962-2660
www.wardsci.com

Chemicals such as caffeine and theophylline can be obtained from most bio-chemical suppliers, such as:

United States Biochemical Corp.
P.O. Box 22400
Cleveland, OH 44122
1-800-321-9322

A pH-neutral mounting medium is difficult to find. Do stay away from Permount, which is acidic. Neutral or near-neutral mounting media are supplied by:

Ward's Natural Science Establishment, Inc.
5100 West Henrietta Road
P.O. Box 92912
Rochester, NY 14692-2660
1-800-962-2660
www.wardsci.com

chapter *12* *Histological Techniques*

Have you ever been awed by the histologist's ability to create a complete set of **serial sections** through an embryo without loosing a single section? This may seem an impossible feat until you learn the relatively easy methods of paraffin sectioning. In this chapter you will have the opportunity to make your own set of serial sections, and in the process, you will be learning a technique that has recently regained the spotlight in embryological studies. With the advent of immunohistochemistry and fluorescence microscopy, developmental biologists have a valuable set of tools, and they need to know the art of tissue preparation to apply them. Regardless of the staining technique used, the basic process is the same: Tissues are first **fixed** to avoid degradation, **embedded** in paraffin, **cut** into serial sections with a rotary microtome, and **stained** to contrast cellular elements. The paraffin technique for tissue preparation is the most commonly used for routine study of tissue. You will be using either chick embryos that you obtained from your previous laboratory exercise (Chapter 11) or other tissue provided by your instructor.

Fixation

Fixation is the first step in any procedure in which tissue is to be preserved for histological study. **Fixatives** kill. They kill the tissue, as well as any bacteria that are present that otherwise would cause the tissue to rot. They also coagulate or cross-link proteins, making them insoluble. All fixatives distort tissue to a certain extent, but in general, proteins and cellular structure are preserved. Normally you choose a fixative containing several ingredients that balance out each other's ill-effects. For example, **alcohol** shrinks tissue and causes excessive hardening. You can counter these effects by adding an acid such as **acetic acid**, which swells tissue and prevents overhardening.

One of the safest fixatives to use, which will not leave any toxic residues behind, is **Carnoy's fixative**, a mixture of alcohol and acetic acid. It is not an ideal fixative. The addition of **formaldehyde**, for example, would give better preservation of cytological detail. Formaldehyde, however, and most other fixing agents leave highly toxic residues that are virtually impossible to remove from instruments and glassware. If you were making whole mounts, where cytological detail is not critical, or you cannot risk contaminating your work area for future live material, Carnoy's is ideal for use.

Carnoy's fixative*

glacial acetic acid	100 ml
100% ethanol	300 ml

Some Carnoy's recipes also include chloroform along with glacial acetic acid and absolute ethanol in the proportions 3:1:6. Chloroform increases the speed of fixation, making it useful for fixing tissues that are difficult to penetrate. Chloroform, however, is extremely hazardous and must be used in a ventilated hood while wearing gloves.

Glacial means 100% acetic acid; vinegar is 5% acetic acid. Ethanol is the type of alcohol that people get arrested for driving under the influence of; hard liquor such as whiskey is about 40%–50% ethanol. Rubbing alcohol is 70% isopropyl alcohol—don't drink it, it's toxic. If you were out in the boondocks with no scientific supplies and found the perfect specimen you needed to preserve, what makeshift fixative would you devise?

A commonly used fixative that is better at preserving structure than Carnoy's is **FAA**, made of **formaldehyde**, **alcohol**, and **acetic acid**. (Formaldehyde is a gas, which is sold in solution as **formalin**, which is about 40% formaldehyde in water.)

Dietrich's FAA**

95% ethyl alcohol	30.0 ml
formalin	10.0 ml
glacial acetic acid	2.0 ml
distilled water	60.0 ml

A fixative that is widely used for embryological material, since it will not overharden yolky material, is **Bouin's fluid**. It has the advantage that tissues may be stored indefinitely in it. Its major disadvantage is that the **picric acid**, though an excellent fixative for lipid-rich tissues that doesn't harden the tissues, also stains the tissue yellow. The yellow color must be removed before other stains can be applied. Usually, this is accomplished after sectioning as the slides descend through the alcohol series. Just leave the slides in 70% alcohol for a longer period of time if the yellow color doesn't disappear using the normal dehydration schedule.

Bouin's Fluid**

picric acid, saturated aqueous***	75.0 ml
formalin	25.0 ml
glacial acetic acid	5.0 ml

---

* Note that under current OSHA guidelines, solutions containing concentrations of alcohol above 24% must be discarded as hazardous waste.

** Under current OSHA guidelines, used fixatives containing formalin must be discarded as hazardous waste, as must solutions containing concentrations of alcohol above 24% .

***Picric acid is sold saturated in water and must be stored that way, since in dry form it is explosive. (The exceedingly destructive explosion in Halifax harbor during World War I was caused by sparks from a careless collision into a munition ship loaded with picric acid.)

To fix your embryo or sample of tissue, place the freshly dissected specimen (it should ideally be no more than 1–2 cm on a side) into a screw-cap vial containing your fixative of choice. For fixation and all steps that follow that require a fluid, the rule of thumb is that the amount of fluid used should be 10× the volume of the tissue. *Label* your vial with your name and what the specimen is, or use a piece of cardstock as your label and place it in the vial with your specimen. *Always use pencil for labeling*—it doesn't come off in the reagents.

The amount of time your specimen is left in fixative should depend on the size of the specimen and the type of fixative. Most specimens should be fixed 6–8 hours, or overnight. If the specimen is smaller than 5 mm on a side, then several hours will suffice. If the fixative severely hardens and shrinks the specimen, as does Carnoy's, then the minimum fixation time should be used. Specimens should not be left in Carnoy's fixative more than 2–3 days. If the fixative is like Bouin's, which is not harsh on the tissue, the tissue may be stored in the fixative for extended periods (months or more) without severe alteration. Tissues should be fixed and stored in the refrigerator at 4°C.

Washing

Following fixation, your tissue sample must be **washed**. This, as well as many of the operations that follow, can be done in the vial. To make a transfer of solution, you can either pour off the old solution into a finger bowl, or you can pipette off the old solution. The new solution is added directly to the vial. This avoids unnecessary handling that can damage the tissue.

Washing is usually done in water. For most fixatives, washing specimens in distilled water for 6–8 hours, or overnight, following fixation is sufficient. If you have chosen Carnoy's as your fixative, then washing in 70% alcohol instead of water is both adequate and faster. You also can use the 70% alcohol as a holding solution.

Following washing, the specimen should be transferred to 70% alcohol and stored at 4°C until it can be processed further. It can remain here for a number of weeks.

Dehydration and Clearing

To prepare the specimen for paraffin embedding, the specimen must be **dehydrated** through a **series of alcohols** up to absolute alcohol (**ethanol** is preferable to methanol, since it is less harsh on the tissues). This removes all the water, which is immiscible with paraffin. By going through a graded series of alcohols, the convection currents that are set up by each transfer are minimized, thereby minimizing harm to the specimen. After the water has been removed, a **clearing agent**, such as **xylene** or **toluene**, which is miscible with both 100% alcohol and paraffin, makes a bridge between the alcohol and paraffin. Toluene is less harsh on tissue than xylene, causing less shrinkage and hardening, and so should be used instead of xylene if possible.

In transferring the specimen through any alcohol solution, the old solution can be poured off (into a finger bowl) and the new solution added. Remember that under current OSHA guidelines, solutions containing concentrations of alcohol above 24% must be discarded as hazardous waste. *Xylene and toluene are toxic.* They should only be used in a ventilated hood (or outside, if necessary) and also must be put into special waste jars and disposed of according to OSHA regulations as a hazardous material.

Making a graded series of alcohol: Always use 95% ethanol (ETOH) to dilute from, rather than 100% ETOH. Absolute (or 100%) ethanol is exceedingly expensive compared to 95%. Here is an easy method for making any percentage of alcohol below 95% that you wish. To make a 70% solution of ETOH, for example, pour 95% ETOH into a graduated cylinder up to the 70 ml line, then add distilled water until the solution reaches the 95-ml line. This gives you 95 ml of 70% ETOH. If you are making a 50% solution, pour 95% ETOH up to the 50-ml line, and add distilled water up to the 95-ml line. To make larger amounts, just use quantities that are multiples of the numbers above.

The following schedule can be used. Times are sufficient for specimens that are no more than 1–2 cm/side. If the sample is larger, longer time periods must be used.

70% alcohol (if starting from water)	1–2 hours
95% alcohol	1–2 hours
100% alcohol (first time)	½–1 hour
100% alcohol (second time)	½–1 hour
Toluene or xylene (first time)*	½–1 hour
Toluene or xylene (second time)	½–1 hour

* *Warning:* If there is any water left in the specimen at this point, it will show as a milkiness around the specimen when it is placed in the clearing agent. The specimen *must* be brought back to 100% alcohol in this case to remove the remaining water.

Special microwaving techniques that save time

Microwaving the specimen during fixation and dehydration cuts down considerably on the time needed in each solution and improves infiltration of the solutions into the tissue. A number of precautions must be taken, however, to make sure that the specimen doesn't overheat, that penetration is uniform, and that toxic exhaust doesn't enter the room. A large beaker of water should be placed in the oven to absorb excess microwave energy. The specimen should not be placed in a "hot" spot of the oven. You can figure out where the hot spots are by placing wet paper towels in the oven and turning it on. Just as the towels begin to dry, turn off the oven. Wherever wet spots remain are good places for placing your specimen vial. Alternatively, the specimen vial can be placed on a rotating table in the oven. To avoid overheating, the oven can be put through cycles of being run on "high" for short periods (7–10 seconds) alternated with being off for 20 seconds. The specimen should not be heated over 70°C—this causes overdenaturation of regions that are relatively impermeable and stain nonuniformly. Putting the vial on a petri dish of ice during processing can prevent overheating.

After 5 minutes of microwaving, a specimen 5 mm × 5 mm × 2 mm can go on to the next step in the fixation or dehydration series. Since microwave ovens vary, you will need to experiment to determine what times and settings work best for your particular oven. *It is exceedingly important that the seals on the oven be checked to make sure that microwave radiation and toxic fumes are not leaking out into the room.* It is safest not to use microwave processing on steps that involve higher alcohols, or clearing agents such as xylene or toluene.

Microwaving can also be used in place of chemical fixation in order to avoid the toxicity of chemical fixatives. Specimens are placed in a saline solution, such as chick Ringer's solution, and microwaved for a period of time, maintaining a temperature of 68–70°C. Small specimens can be microwaved for as little as 6 minutes. Again, you will need to experiment to determine the times and settings that work best for your particular oven.

Paraffin Infiltration

From the second change of toluene, the specimen is to be moved through several changes of melted paraffin. The paraffin you use will have a specified melting point. A melting point of 54°C is typical and is suitable for sections that will be cut at 8 μm or thicker. Higher melting points indicate harder paraffin that can be cut into thinner sections.

Typically, the melted paraffin is kept in a paraffin oven or paraffin bath that will maintain the paraffin at a constant temperature. The specimen can be transferred to a cage specifically designed for infiltration. Be sure to include a label in the cage, along with the specimen, that distinguishes your sample from other samples in the oven. The cage can then be transferred from one bath of melted paraffin to another using forceps or a string tied to the cage. Alternatively, deep-welled spot dishes containing melted paraffin can be used in a paraffin oven and the specimen transferred from one well to the next. In this method, use a narrow piece of cardstock that you hold with forceps to lift the specimen for each transfer—this avoids the harm to the specimen that would occur if a metal instrument such as forceps were used to pick up the specimen.

The paraffin should be kept above its melting point during this process, but not hotter than necessary, since heat severely shrinks and hardens tissue. Use two to three changes of paraffin, ½–1 hour each. Under ideal circumstances, you would be able to infiltrate in a vacuum oven—this gives good infiltration and shortens the time needed in hot paraffin.

Paraffin Molds

One of the most successful molds for embedding is a paper box made such that the seams are watertight. Such a box can be made to any size appropriate for your specimen. It also has the advantage, over molds with rigid sides, of being pulled inward along with the paraffin as the paraffin cools and shrinks. This prevents a deep indentation from forming in the center of the block. More expensive manufactured molds made of plastic or metal can also be used, and these can be bought from supply houses.

I much prefer paper boxes to any manufactured mold. A paper embedding box should be made of heavyweight paper. It is made most easily by folding it over a block of wood that has one face that matches the size that you need. For a specimen approximately 5 mm^3, use a rectangular sheet approximately 6 cm × 10 cm. Place this on a block of wood with a face approximately 2 cm × 2 cm (Figure 12.1A). First fold the long sides of the paper down along the wood (Figure 12.1B). Holding these in place with one hand, use the other hand to fold one of the remaining sides down (Figure 12.1C). Crease the two corners at a 45° angle (Figure 12.1D). This creates flaps, which are then folded inward,

Figure 12.1
Making a paper embedding box. (A) Place a block of wood in the center of a rectangular piece of paper, approximately 6 cm × 10 cm. (B) Fold the long sides of the paper down alongside the wood. (C) Hold the two long sides against the wood with one hand, and use the other hand to fold one of the remaining sides down alongside the wood. (D) Crease the two corners at a 45° angle and fold each flap inward. This creates a tab. (E) Fold the tab downward. (F) Repeat steps C–E on the remaining side.

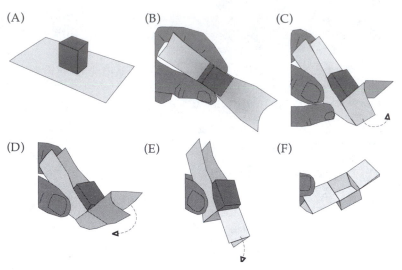

and a tab that is folded down (Figure 12.1E). Do the same to the remaining side (Figure 12.1F).

Paraffin Embedding

Fill the embedding mold with melted paraffin and transfer the specimen from melted paraffin immediately to the mold. The specimen can be transferred using a small strip of cardstock as a spatula. This prevents damage to the specimen. Position the specimen in the mold using a hot needle, heated in the flame of an alcohol lamp. The orientation of the specimen should be appropriate for later sectioning of the block. If using a paper box, this orientation can be marked on the tab of the box, along with information on what the specimen is and any other information necessary to identify it later.

After a film of solidifying paraffin has formed over the surface of the paraffin, the mold should be transferred to a bowl of cool (but not ice-cold) water (water that is too cold can cause the paraffin to crack). The water causes the paraffin to cool more quickly, making for smaller crystal formation within the paraffin as it solidifies and hence less distortion of the specimen. Once the block is hard, remove it from the water and store in the refrigerator. You should not try mounting and cutting the block until it has cooled for 6–8 hours, or overnight. This improves the cutting process and is one of those magic tricks one learns but can't explain.

Mounting and Trimming Paraffin Blocks

The paraffin block is to be mounted onto a piece of metal, synthetic fiber, or wood, all of which can be referred to as "**chucks**," which fit into the chuck holder of the microtome. When mounting the paraffin block onto the chuck, the chuck must be held securely. A heavy piece of wood with holes cut into it or even a chunk of clay into which you press the chuck are suitable holding devices.

Rough trimming Remove the paraffin block from its mold, keeping track of the orientation of the specimen within it. Trim away excess paraffin using a safety razor blade. This should be done on a protective piece of wood or glass. Do not trim too close to the specimen. Leave at least 5 millimeters of paraffin on all sides and more than this on the bottom of the block. Paraffin scraps from trimming can be remelted, filtered, and reused. Paraffin improves with reheating, since impurities are volatilized off with each heating.

Mounting the block   Using the flame from an alcohol lamp, melt some shavings of paraffin on the blade of a metal spatula and pour the melted paraffin onto the surface of the chuck. In this way, you are to build up a layer (at least 1 mm thick) of paraffin that is firmly attached to the chuck. Heat the spatula again and use it to melt the top of this layer of paraffin. Then heat both sides of the spatula blade, hold it between the bottom of the paraffin block and the paraffin on the chuck so that it melts both surfaces of paraffin, and slide it away as you press the paraffin block against the chuck. Heat the spatula blade again, and use it to melt some paraffin on the four vertical sides of the paraffin block to ensure that it is attached firmly to the chuck. Set aside until the paraffin is firm.

Fine trimming   You now need to trim the face of the block so that it is in the shape of a trapezoid, ⏢ with the top and bottom sides being parallel.

The slanted sides help the sections that are cut to stick together into a ribbon rather than sticking to the knife. If the top and bottom sides are parallel, then the ribbon that is cut will be straight. There should be about 2 millimeters of paraffin left around the specimen on all sides.

Sectioning

You must be familiar with the controls on the **rotary microtome** before you can proceed with **sectioning** (Figure 12.2). Be sure you have been checked out by an instructor and received permission to go on to the next step.

The wheel of the microtome should be in the **locked position**. Clamp the chuck containing your paraffin block into the **chuck holder** on the microtome. Using the control screws, align the chuck holder so that the face of the block is absolutely vertical. The block, when you look at it straight on, should look like the picture shown above, with the short side of the trapezoid uppermost.

Insert the **microtome knife** into the **knife holder,** and tighten the clamping screws. This should always be done with great care. These knives are exceedingly sharp and dangerous. If you should drop one, let it go—*never try to catch it*. (You could lose a finger by attempting such a catch.) These knives are also very expensive, so it is worth taking care of them.

Advance the knife toward the block. This is done by unlocking the wheel, lowering the paraffin block until it is on a level with the knife-edge, and then carefully moving the **knife carriage** forward until the knife is just in front of the paraffin block. Lock the knife carriage in place.

Before cutting sections, check the thickness setting. Paraffin sections are typically cut at between 4 and 12 µm in thickness. For routine sectioning, set the thickness to between 8 and 10 µm.

To section, turn the handle of the microtome, which advances the block toward the knife. On each rotation of the wheel, the block is moved forward the number of microns that have been set for sectioning thickness. Turn the wheel in a smooth slow motion to ensure getting the best sections. When sections begin to come off the block, the first several may be incomplete. These can be brushed

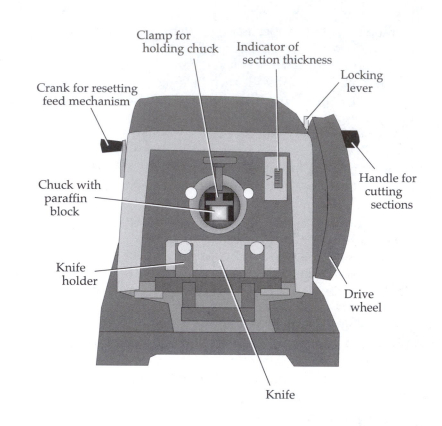

Clamp for
holding chuck

Indicator of
section thickness

Locking
lever

Crank for resetting
feed mechanism

Chuck with
paraffin
block

Handle for
cutting
sections

Knife
holder

Drive
wheel

Knife

Figure 12.2
A rotary microtome
used for cutting paraffin
sections.

away with a **camel's-hair paintbrush**—no other instrument should ever be used, since these would damage the knife-edge. And *never* allow your fingers to get anywhere near the knife-edge.

Once a **ribbon** of sections begins to form, the end of the ribbon can be lifted with a moistened camel's-hair paintbrush. Do not pull on the ribbon; simply support its end with the brush. When you have a suitable length of ribbon (no longer than the length of your ribbon box), take a second moistened camel's-hair paintbrush, and use it to lift the attached end of the ribbon away from the knife-edge. Place the ribbon carefully into your **ribbon box** (a flat stationery box does fine). Cover the box so that air currents don't disturb your ribbon. Keep sectioning until you have all the sections you need. If you are making a set of serial sections, keep each ribbon you cut in order, laying them down so that the last section of each ribbon is on the right, and each successive ribbon is below the previous ribbon.

Retrimming the block   As you section, you may need to retrim the paraffin block if the block face becomes too large or if the ribbon is not straight. Remember, the ribbon will not be straight unless the top and bottom sides of the block are precisely parallel to one another. When retrimming the block, the knife *must* first be removed from the knife carriage. Do this without changing the position of the carriage. Retrimming with a safety razor blade may now be done on the paraffin block in place without removing the chuck from the chuck holder.

After retrimming the block, replace the knife, and back the paraffin block off a slight distance (an eighth of a turn) from the knife using the handle for this—it is on the opposite side from the sectioning wheel. This will avoid taking too thick of a section on the first rotation of the wheel.

Cleaning the knife   When you have finished sectioning, the knife should be cleaned of any paraffin that is sticking to it. First attach the knife handle to the knife. This will allow you to hold the knife without danger. Place a few drops of xylene or toluene on the edge of the knife, and with a Kimwipe®, wipe very carefully, moving from the flat of the knife off the edge of the knife, as you would to brush crumbs off a table. *Never* wipe along the edge of the knife or into the edge. Remove the knife handle and place the knife in its holder (sharp edge down).

Mounting Sections

No matter how carefully you cut your sections, the cutting process compresses the sections, and part of the mounting procedure is to **expand** the sections before they adhere to the slide. This can be done by floating the sections on warm water (5–10°C below the melting point of the paraffin). Some histologists float the sections in a water bath and then pick the expanded sections up on a slide. I find it easier to put water on the slide itself, float the sections on the water, warm the slide on a warming tray, and then pour off the water after the sections are expanded. This second procedure is the one we will use.

Select the number of slides you think you will need. Even though these have already been factory-cleaned, they should be cleaned again by at least rubbing them with a dry Kimwipe®. Then, using a diamond marker, etch into one end of the slide the information that is necessary to identify it (usually, specimen or experiment number and the number of the slide in the series that you are making).

An **adhesive** must now be applied to the slide so that the specimen sections will adhere properly and won't fall off during further processing. If you are not trying to identify specific proteins in your staining procedure, then the best adhesive to use is Mayer's albumen. This can be made easily and stored indefinitely. Mayer's albumen can even be left out at room temperature, but it is better to store it in the refrigerator.

Mayer's albumen adhesive:
> Equal parts:
>   glycerin
>   egg white (shaken first to break
>   it up and make it slightly frothy)

> *Add a few crystals of thymol as a preservative. Store in a dropper bottle, in the refrigerator.*

For use   Add a drop of adhesive to each slide. Using a Kimwipe®, smear the drop across the area of the slide where you will be placing sections. You should see a shiny film of adhesive when you tilt the slide into the light. It is important not to apply too much adhesive, since it will stain during the staining procedure and will cause too much background staining if applied too thickly.

Now add a dropper full of boiled distilled water from a dropper bottle. This water should have been recently boiled and cooled. The boiling removes air from the water and helps to prevent air bubbles from forming underneath your sections.

You can now place lengths of paraffin sections onto the slide. Using a scalpel or safety razor blade, cut your paraffin ribbons into lengths. These should be no longer than half the length of the coverslip that you'll be using.

Mounting the ribbons   Using two camel's-hair paintbrushes to hold the two ends of each piece of ribbon, transfer pieces of ribbon to the water on the slide. Ribbons of serial sections must be placed in order, one under the other, until the width of the slide is filled (Figure 12.3). It is best if the ribbons are touching one another along their length so that they are rafted together. This makes it easier to manipulate the ribbons later.

Place the slide on a **slide warming tray** that is set at a temperature that is 5–10°C below the melting point of the paraffin. Usually this is a temperature that feels very warm to the touch but not hot enough to burn you. As the water warms on the slide, the sections will expand, greatly increasing the length of the ribbons.

When the ribbons look completely expanded (this takes about 1 minute), remove the slide from the warming tray. Tip one corner of the slide against absorbent paper, such as a Kimwipe®, to wick off the water without disturbing the ribbons. You need to remove only as much water as will pour off. This leaves a film of water behind that allows the repositioning of any ribbons that have become displaced.

Using a dissecting needle, reposition the ribbons so that they are arranged as you wish on the slide. Place the slide back on the warming tray for about 10 seconds. This will help flatten any slight ripple you might have caused when repositioning the ribbons. It is extremely important that you do not melt the paraffin, since this will seriously distort your sections. Remove the slide from the tray and place in a slide holder.

Drying slides   The slides must now dry for at least 6–8 hours (or overnight) before staining. If any water remains prior to staining, the sections will not stick properly to the slide. It is best to dry the slides in a slide-drying oven (set on warm) or in a jar that contains desiccant.

Staining

During the processing of the slides, the general scheme is that first the paraffin is removed from the sections using xylene or toluene, since the paraffin prevents staining of the tissues. The slides are then processed down to water, since most staining solutions are water-based. This requires a graded series of alcohols to prevent severe convection currents that would damage the tissues. Following staining, the slides are dehydrated again through a graded series of alcohol, cleared in xylene or toluene, and coverslips are applied using a plastic mounting medium that is miscible with the clearing agent.

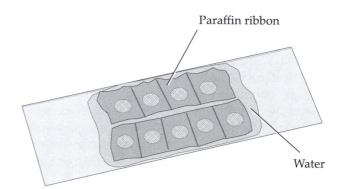

Paraffin ribbon

Water

Figure 12.3
Ribbons of paraffin sections floating on water on a slide.

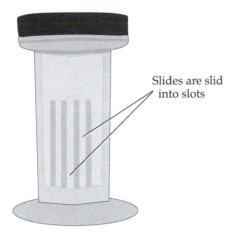

Slides are slid into slots

Figure 12.4
A Coplin jar used in staining slides. Slides are placed in slots between glass ridges to keep them from touching one another.

For clearing agents, it is better to use toluene than xylene since it causes less shrinkage. For alcohol, it is better to use ethanol than methanol since it is gentler on tissue and is nontoxic.

Setting up the staining series   There are a variety of types of staining dishes that can handle a few slides up to many slides at a time. For up to nine slides, **Coplin jars** (Figure 12.4) are used (place every other slide at an angle so that it is sharing a slot with the slide in front of it on one side and sharing a slot with the slide behind it on its other side). It is best, however, to process no more than five at a time; each slide is slid into a pair of slots and is separated from the next slide by a pair of glass ridges. *This ensures that no slide touches another slide*. A distinct advantage to Coplin jars is that most have screw tops, which will prevent evaporation of solutions between use.

The Coplin jars should be set up and labeled prior to the beginning of staining. For a typical hematoxylin, eosin, and alcian blue staining process, set up the following series of Coplin jars and label both the jars and their tops. Use label tape and always label in pencil since toluene (and xylene) as well as alcohol typically dissolve the ink of other markers. On a jar that is in the "down" series, a downward pointing arrow can be used to indicate this, and on a jar that is in the "up" series, an arrow pointing upward can be used. In the down series, there are two jars of toluene (or xylene) to ensure that all paraffin is removed. On the up series, there are two jars of 100% ETOH to ensure that all water is removed from the sections. These are the two most critical steps in the series for determining success.

Coplin jars for a hematoxylin, eosin, and alcian blue staining series

- Toluene (or xylene) I — ↓
- Toluene (or xylene) II — ↓
- 100% ETOH — ↓
- 95% ETOH — ↓
- 70% ETOH — ↓
- Distilled water — I
- Alcian blue
- Distilled water — II

- Hematoxylin
- Tap water
- 70% ETOH — ↑
- Alcoholic eosin
- 95% ETOH — ↑
- 100% ETOH I — ↑
- 100% ETOH II — ↑
- Toluene — ↑

Using the hematoxylin, eosin, and alcian blue staining series

Hematoxylin and eosin (H&E) are the most commonly used general nuclear and cytoplasmic stains. Hematoxylin (a basic dye) stains acidic components, primarily nucleic acids, a dark blue and is therefore used as a nuclear stain. Eosin (an acidic dye) has an affinity for cytoplasmic elements. Its yellowish-red color makes it an ideal counterstain to hematoxylin. Alcian blue stains glycosaminoglycans (e.g., mucus, chondroitin sulfate of cartilage) and is a simple addition to the H&E series that adds useful information.

Reasons for the order of things   Because alcian blue is used at a very acidic pH, it must be used prior to hematoxylin in the staining series to avoid removing any bound hematoxylin. A tap-water bath is placed after hematoxylin as a "bluing agent." Tap water is slightly basic, and a basic solution is needed to change the color of the bound hematoxylin from a rusty brownish color to a dark, almost black, blue. Moving up from eosin should be done as quickly as possible to avoid losing too much eosin from the tissues. Processing can be speeded up by gently dipping the slides in and out of the solutions to increase the speed of diffusion. This must be done *gently* to avoid detaching the sections from the slide.

Filtering stains before use   A precipitate forms in most stains, requiring that they be filtered before use. Hematoxylin will need to be filtered if it has been sitting for more than a day. Alcian blue should also be filtered but doesn't need filtering as frequently—once a week should be sufficient. Alcoholic eosin forms little precipitate. Simply check for a precipitate and filter if necessary. (Unbleached coffee filters make inexpensive filters that are sufficient for removing the precipitate from stains.)

Remember that toluene and xylene are toxic, and these should only be used where there is good ventilation. Use in a fume hood (or handle them outside!).

Staining schedule

Toluene (or xylene) I — ↓	20 min.
Toluene (or xylene) II — ↓	20 min.
100% ETOH — ↓	5 min.
95% ETOH — ↓	5 min.
70% ETOH — ↓	5 min.
Distilled water — I	5 min.
Alcian blue (filter before use)	10 min.
Distilled water — II	5 min.
Hematoxylin (filter before use)	2 min.
Tap water — several changes	1 min. each
70% ETOH — ↑	5 min.
Alcoholic eosin	5 min.
95% ETOH — ↑	1 min. dipping
100% ETOH I — ↑	1 min. dipping
100% ETOH II — ↑	1 min. dipping
Toluene (or xylene)* — ↑	2 min. and hold

*On transferring your slides to the final toluene (or xylene), if you notice any white clouding of the fluid, this means that there is still water on the slide and in the tissues. You *must* immediately go back to 100% ethanol. If you don't, there will be water droplets in your preparation that will make your slide look like salad dressing.

Formulary for staining solutions

Alcoholic eosin

eosin Y, saturated solution in 95% ETOH (approx. 0.5 gm/100 ml)	10 ml
95% ethanol	45 ml

Alcian blue

alcian blue	1 gm	*Filter before use.*
distilled water	100 ml	
glacial acetic acid	3 ml	
thymol (to prevent mold)	a crystal	

Hematoxylin

There is true art to making hematoxylin solutions. You can buy them already made, or you can make them yourself. For any hematoxylin solution, it is important to know the shelf life of the solution. Some will keep for years (Delafield's and Erhlich's), and some last only a month or two (Harris's).

Ehrlich's hematoxylin

hematoxylin	2 gm	*Ripen for 6–8 weeks.*
ammonium alum		*Add 10 ml glacial acetic*
$(NH_4Al(SO_4)_2 \cdot 12H_2O)$	3 gm	*acid. Keeps for years.*
100% ethanol	100 ml	
glycerin	100 ml	
distilled water	100 ml	

Mounting Coverslips

You are now ready to put a **coverslip** on your slide using a **mounting medium**. Since you are going from toluene (or xylene) to this mounting medium, this step should be done in a fume hood.

The mounting medium you use should be a neutral mounting medium such as Histoclad. Cheaper mounting media, such as Permount, are acidic. They cause the stain to fade and oxidize over time, making a cloudy ring around the edge of the slide that gradually works its way to the middle. Any slide that needs to be permanent is worth the extra cost of a neutral mounting medium.

The hardest part of mounting is devising a method to lower the coverslip without introducing air bubbles into your preparation. Once you start the mounting procedure, work with speed. Do not let your sections dry out. They will do so very quickly. The amount of mounting medium you should put on the slide depends on the length of the coverslip you are using. Use an eyedropper to put a pencil's width of medium down the length of the slide to match the length of the coverslip. Hold one end of the coverslip with forceps, and lower the other end onto the slide, holding it in place with a dissecting needle. Slowly lower the rest of the coverslip. If an air bubble is introduced, it will usually move to the edge as you lower the coverslip. Once the coverslip is in place, very slight pressure on the top of the coverslip using the end of a pencil or dowel (you can use the handle end of your dissecting needle) will push out any bubbles on the edges. This *must* be done gently to avoid damaging the sections. In most cases, bubbles should just be left where they are. A few bubbles are better than smashed sections.

If you have too much mounting medium so that it is oozing out from under the coverslip, don't touch it. After it dries for a week, excess mountant can be removed with a razor blade (carefully).

Allow your slides to dry flat in a dust-free environment (such as a small cardboard box used for stationery or a flat cardboard slide holder). These should be left at room temperature for several days before observing them under the microscope. After a week, once they are thoroughly dry, they can be stored on their sides in regular slide boxes.

Accompanying Materials

Tyler, M. S. and R. N. Kozlowski. 2000. *Vade Mecum: An Interactive Guide to Developmental Biology*. Sinauer Associates, Sunderland, MA. "Histotechniques." This chapter of the CD illustrates, with movies and still pictures, how to do each of the procedures described here.

Selected Bibliography

Bird, M. 1995. *The Town That Died: A Chronicle of the Halifax Disaster*. Nimbus Publishing, Ltd., Halifax, N.S. A history of the explosion caused by picric acid loaded on a munition ship that caught fire in Halifax harbor during World War I, the worst human-made explosion before the nuclear explosions of WW II.

Boon, M. E. and L. P. Kok, (eds.). 1992. *Microwave Cookbook for Microscopists*. Coulomb Press, Leyden, Leiden, The Netherlands. This gives many of the details that will be useful in applying microwave techniques to your histology protocol.

Gleiberman, A. S., N. G. Fedtsova and M. G. Rosenfeld. 1999. Tissue interactions in the induction of anterior pituitary: role of the ventral diencephalon, mesenchyme and notochord. *Dev. Biol.* 213: 340–353. This research paper is a fine example of the use of paraffin histological techniques in immunohistochemistry.

Gu, J. (ed.). 1997. *Analytical Morphology: Theory, Applications and Protocols*. Eaton Publishing Co., Natick, MA. A valuable book to the sophisticated microscopist who needs to delve into immunohistochemistry and fluorescence microscopy. It also has an excellent discussion of the use of a microwave oven to speed up the processing of tissues.

Humason, G. L. 1979. *Animal Tissue Techniques*, 4th Ed. W.H. Freeman, San Francisco. This is a classic, and still very useful, handbook on histological techniques. It gives very complete instructions on each step in the processing of tissue, as well as recipes for fixatives and stains, and explains what each is specific for.

Login, G. R. and A. M. Dvorak. 1994. Methods of microwave fixation for microscopy. A review of research and clinical applications: 1970–1992. In W. Graumann, Z. Lojda, A. G. E. Pearse and T. H. Schiebler (eds.), *Prog. Histochem. Cytochem.* 27: 1–127. This guides you through steps needed to adjust the microwave protocol for processing tissues to the particular microwave oven you are using.

Presnell, J. K., G. L. Humason and M. Schreibman. 1997. *Humason's Animal Tissue Techniques*, 5th Ed. Johns Hopkins Univ. Press, Baltimore. This is a welcome reissuing of Humason's classic volume.

Suppliers

A large biological supply house, such as:

Fisher Scientific
585 Alpha Dr.
Pittsburgh, PA 15238
1-800-766-7000
www.fishersci.com

A full line of histological supplies, including reagents, stains, staining dishes, slides, coverslips, embedding media, mounting media, microtomes and microtome supplies, slide warmers, and slide boxes.

Ward's Natural Science Establishment, Inc.
P.O. Box 92912
5100 West Henrietta Road
Rochester, NY 14692-9012
1-800-962-2660
www.wardsci.com

Ward's offers an economical line of student microtomes for cutting single sections at a time, not suitable for serial sectioning. They also sell a selection of histological supplies and neutral mounting media.

chapter 13 *Planarian Regeneration*

Regeneration, the process of molding tissues into an exact replica of missing parts, is a mystery worthy of many hours of armchair biology, punctuated by periods of clever experimentation. It is a field for adventurers, where amateurs can have the advantage, since their minds are still free of prejudices. When you venture into the field, be wary of simple solutions. Doubt all theories and devise your own. Test theories through experiments, and let your organism serve up the answers. The oddities that appear to defy explanation may be the clues that everyone else threw away in frustration. Pocket these to mull over in quiet hours. And above all, don't let confusion cloud your admiration for the phenomenon: regeneration is truly astonishing. Remember also that the ultimate prize would be understanding regeneration in enough detail to be able to trick a normally nonregenerating system, such as the human hand, into regenerating parts tragically lost in accidents.

Planarian flatworms offer an ideal system for regeneration experiments. Mutilation of an organism is never a game; however, in the planarian, cutting an animal in half is doing no more than the animal does to itself. When reproduction is in order, the animal adheres tightly to the substrate with its posterior end, pulls forward with its anterior end, and literally tears itself in two. The two halves then regenerate their lost half. This process of asexual reproduction, called **fission**, produces genetically identical individuals—true clones. Many times an experimenter has returned to a control animal after a few days to find the single control is now two. You may see it happening in your own dishes, so watch carefully.

Tearing oneself in half may seem a peculiar mode of reproduction, but it is not the only mode available to the planarian. Its sexual form of reproduction, however, may seem no less peculiar than its asexual mode. Planarians are **hermaphrodites**, containing both male and female reproductive organs within a single individual. Among many planarians, sexual reproduction is rare; it may occur only seasonally, and in some races it doesn't occur at all. When it does, fertilization is internal, and the paired organisms each inseminate the other. Sperm are introduced into the mate using a copulatory bulb inserted into a genital pore. Sperm travel up the oviducts and are held in storage near the ovaries until the eggs are ovulated. Eggs traveling down the oviducts are fertilized by the stored sperm. Yolk cells are then added around the eggs by specialized regions of the oviduct. A cocoon, or egg capsule, is secreted to surround a group of eggs, and these capsules are extruded through the genital pore and attached to solid substrata. Though the possibility exists for an animal to fertilize its own eggs, this probably rarely, if ever, occurs.

Collecting and Maintaining Planarians

Planarians are free-living, freshwater flatworms, classified as **triclad turbellarians**. The common genera—*Planaria, Dugesia, Polycelis,* and *Phagocata*—include many species that have been used for regeneration studies. Two in particular, *Dugesia dorotocephala* and *D. tigrina* (brown planarian), are large enough to be satisfying for classroom studies and are commonly available from biological supply companies.

If you are braving the wilds to collect your own planarians, there are several hunting methods you can try. Planarians are often found in abundance on the undersurface of stones or wood in the swiftly flowing water below dams, in spring-fed streams, along the shores of lakes, and even in marshes. If the water is not frozen, collections can be made even in midwinter. You may find the animals simply by turning over stones and logs and seeing them adhering to the undersides. You can dislodge the animals using a stream of water from a pipette directed into a bucket or using a soft paintbrush to transfer them. Sometimes you can catch a number of planarians by placing small pieces of fresh liver at the edges of flowing water, waiting 15–20 minutes, and then inspecting the pieces. Planarians, attracted to the liver by chemotaxis, should be feeding on its shaded underside.

Another method, less delicate, is to bring in masses of submerged vegetation from streams or lakes. The vegetation should be kept covered with water until it starts to decay. If planarians are present, they will collect at the surface of the water and along the sides of the container, where they can be picked off with a soft paintbrush.

Planarians in the laboratory should be maintained in very clean conditions. They should be kept in water in enamel pans or glass finger bowls that are kept in darkened conditions. The water cannot be tap water, since the chlorine in treated water is toxic to these sensitive animals. Spring or well water, or water from the collecting site, should be used, and the water should be changed one or two times a week. Each time the water is changed, the dishes should be either changed or rubbed down to remove the accumulated "scuzz" that builds up from the copious mucus these animals secrete.

If the animals must be maintained for a long period prior to the experiments, they should be fed once or twice a week. Once a week is sufficient for maintaining the cultures; twice a week is only necessary if you want to increase the stock quickly. The animals can be fed on raw beef liver or egg yolks. Leave the food in the dishes no more than 2–3 hours, then remove it and change the water. This will prevent bacteria from fouling the water.

Prior to your regeneration experiments, the animals should be starved for a week. This will empty the gut of food and will avoid bacterial contamination during the recovery period. Animals will not be fed again until regeneration is complete.

Observations of Normal Anatomy and Behavior

Use a soft paintbrush to transfer animals from the large enamel pans or finger bowls to a petri dish containing well or spring water. You will be keeping one animal as a control. The others will be used for regeneration experiments.

Anatomy

Observe the intact animal under the dissecting microscope, and learn its anatomy using Figure 13.1. This knowledge will be invaluable to you when you start analyzing your regenerates. You will be able to discern a great deal more anatomy by making a planarian "squeeze." Put a ring of petroleum jelly on a coverslip, place the planarian in the center with a drop of water, then press a second coverslip on top. Apply enough pressure to squeeze the animal flat without damaging it. Place this on a slide and observe it under both the dissecting microscope and the low power of your compound microscope. By turning the preparation over, you can observe both the dorsal and ventral surfaces. Make sure you adjust the light in a variety of ways; certain structures will be seen better with direct light, others with transmitted light.

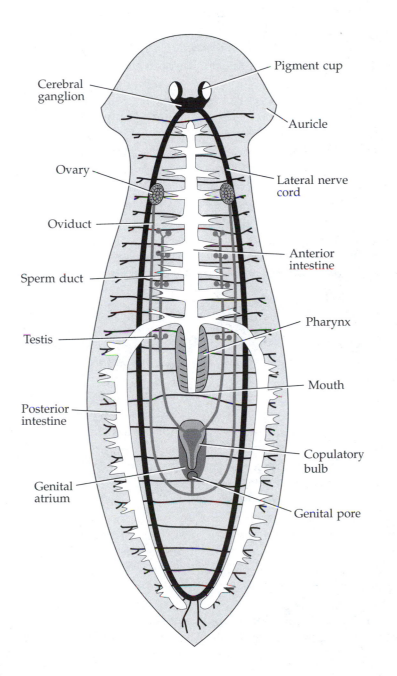

Figure 13.1
Schematic diagram of planarian anatomy. Dorsal view.

Notice that the dorsal surface is pigmented and more darkly colored than the ventral surface. When doing regeneration experiments, this will be important for keeping track of dorsal and ventral surfaces of a piece you might be grafting. Look for the **gut**; this should be the most obvious structure, almost filling the body. It lies dorsal to most other organs. In the midline and opening ventrally is the extrusible **pharynx**. Turn the preparation so that you are looking at the ventral surface. Notice that the pharynx is a highly muscular structure and will look quite dense. The pharynx is used to suck food—fish debris, small prey, even other planarians—into the gut. (If a planarian fissions in two and then cannibalizes its other half, does it count as self-destruction?)

The food, once it enters the mouth, travels through the gut, which is **tripartite** (hence the name triclad). The gut, with a single **anterior branch**, two **posterior branches**, and numerous **subbranches**, extends to almost all parts of the body. Food is processed down to the size of small particles in the gut. These particles then are phagocytized by gut epithelial cells for their final digestion. Undigested food in the gut exits by the same door it entered, the mouth; there is no anus. The animal first fills its gut with water and then forces it out by muscular contraction to irrigate the entire gut.

Look from the ventral side for the bilobed **brain** from which extend two **lateral nerve cords** that have numerous **transverse connections**. The nervous tissue will be whitish. Attached to the brain anteriorly are the **photoreceptors**, which are clustered together into two clumps, each shaded by a cup of **pigmented cells**. These paired **eyespots** give the animal its "cute" cross-eyed look. The pigmented cup ensures that the photoreceptors are receiving light only from one side. When you are observing regenerates to determine whether head regeneration has taken place, the presence of these eyespots will be one of your clearest indicators of new head formation.

Look at your animal from the ventral side for the presence of reproductive organs. You will be lucky if you see any, not because they are invisible, but because in many planarians these develop only during seasons of sexual reproduction—usually spring and summer—and regress for the rest of the year. If the reproductive organs are present, you should see paired **ovaries** anteriorly, with **oviducts** that lead down to the **genital atrium**. **Testes** are multiple and appear along the length of the body. They empty into **sperm ducts** that are connected to a **copulatory bulb** in the genital atrium.

Behavior

After you have observed the internal anatomy, release your animal from its glass sandwich, place it in your dish of water, and let it recover. Use your other planarians to observe some of the basic behavior of these animals. Record your observations, for these will be used as a standard against which you will evaluate your regenerates.

Look carefully at the animal. How does it move? Does it glide, swim, or crawl? If you flip it over, can it move on its dorsal surface? Can the animal right itself? The animal has a **ciliated ventral surface**, and this will allow the animal to glide along surfaces, but far more frequently, it uses its **longitudinal** and **circular musculature** for movement, crawling along a surface of mucus that it secretes from its ventral surface.

Test for various senses in the animal. How does it respond to touch? Try poking and stroking the animal at its anterior and posterior ends. Does the animal respond differently depending on where it is touched? The animal does have **tactile receptors**, which are concentrated in the head.

Your animal should also exhibit chemoreception (as you already know if you caught them on a liver-baited fishing line). The **chemoreceptors** are concentrated in ciliated grooves located on the auricles. These **auricles** are the earlike extensions on the head (adding another measure of "cuteness" to the animal). If you place a small drop of liver juice in one part of the dish and create a slight current with a pipette toward the animal, the animal may respond for you, demonstrating its chemoreception. Record its responses.

You may have already noticed that planarians tend to aggregate in any shaded region of the dish. They are demonstrating **negative phototaxis**, responding to the light that they sense with their photoreceptors by moving away from it. You can determine the angle of light that each eyespot is able to perceive by shading your dish and shining light from different directions. What advantage is there to the animal of being negatively phototactic?

Once you have observed and recorded all the behavioral responses you can think of testing, move on to the regeneration studies.

The Regeneration Process: Patterns and Theories of Regeneration

Regeneration experiments must be carefully planned so that they test specific theories. Before you start, it is important that you know something about planarian regeneration, know the models that are thought to explain it, and dabble in creating your own. Study the next section well before you make your first cut.

The regeneration process

The first step in regeneration is wound closure and wound healing. When you make your cuts, you should be able to observe these processes by watching the cut surface under the dissecting microscope. First the wound will close by muscular contractions of the body wall. This will take about 10 minutes. The epithelium then heals over the wound, actively spreading, which takes an additional 20 minutes. Once the wound has healed, a **blastema** begins to form. The blastema is an accumulation of undifferentiated cells that eventually will differentiate into the missing parts. The cells of the blastema are called **neoblasts**. They are embryonic-like cells that are found throughout the body, held in reserve for just this task. When a cut occurs, the neoblasts in the immediate area coalesce to form the blastema. (There is evidence that neoblasts can also migrate in from more distant regions, but they probably do so only in unusual circumstances when local neoblasts have been destroyed.) The cells at the base of the blastema are highly mitotic, while those in the rest of the blastema are not. In animals held at 22–24°C, the blastema usually forms within 1–2 days and should be clearly visible to you by 3–4 days as an unpigmented region. The timing of events is highly dependent upon temperature, and it is slower at lower temperatures. At 22–24°C, differentiation proceeds quickly, and by 4–6 days you should see differentiated structures in the regenerating region; by 2–3 weeks, regeneration should be complete, with regenerates having reestablished normal body proportions.

This type of regeneration, in which the lost parts are built anew from a population of undifferentiated cells, is called **epimorphosis**. This is in contrast to a process called **morphallaxis**, in which differentiated cells are remodeled into the new parts. Planarians undoubtedly undergo epimorphosis, but this should not blind us to the possibility that morphallaxis may also be contributing. Some research indicates that morphallaxis may even be playing a major role in the process (Chandebois, 1984).

Pattern formation and positional information

It has been known since the 1700s that when a planarian is cut in half, or into even smaller portions, each piece is able to accurately regenerate an entire animal. Even early workers in the field, such as Charles Manor Child and Thomas Hunt Morgan (two names that should be known to any budding developmental biologist), realized that there must be a system of information that specifies what should be regenerated. How is this information encoded? Child and Morgan considered that **gradients** of substances must be involved, but substantiating this was difficult. Even now, almost a century later, we still don't have any specific answers. What we do have is a set of terms that define what we're looking for: **pattern formation** refers to the process of creating specific patterns and shapes out of tissues, and **positional information** refers to the information necessary to create these patterns and shapes (see Wolpert, 1978).

The models proposed to explain pattern formation and positional information are too many to examine here, so we will concentrate on two models, a **diffusion gradient model** and the **polar coordinate model**. These should provide you with ammunition to start inventing other models on your own—it's open season, and your models can be as good as anyone's.

Diffusion gradient model

A number of authors have suggested a diffusion gradient model for positional information in regenerating systems (see Wolpert, 1978). Whether or not such a model is applicable to planarian regeneration is something you can test in your experiments.

A diffusion gradient model assumes that positional information is established by a gradient of a chemical substance (**morphogen**). Cells would know their positions by sensing concentration levels of the morphogen, and they would know their polarity (which way is "up") by sensing the direction of the gradient. Such a model divides the organism into a number of regions, usually along the anterior-posterior axis (represented in Figure 13.2 as A–E; the number of regions was chosen arbitrarily). The regions represent different threshold levels of the morphogen. Cells are thought to respond by forming different structures at different thresholds.

The model predicts a number of outcomes that you can test. For example, if the animal were cut in half, the model predicts that the level of morphogen at the cut surface of each piece would tell the cells forming the blastema where they were in relation to the whole and their polarity. The anterior half would know to regenerate a tail end, and the posterior half would know to regenerate a head end (see Figure 13.4A). This is fairly straightforward. It is far more challenging to design experiments that "confound" and really test the model. For example, suppose, as shown in Figure 13.3C, only section B were removed and the remaining two pieces reannealed? What would happen? The model predicts that there would be remodeling of tissues posterior to the cut due to a change in gradient

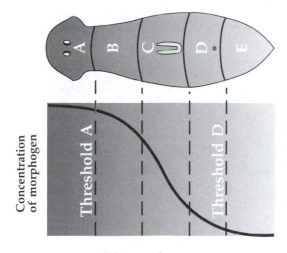

Figure 13.2
Schematic diagram of a diffusion gradient model of positional information as it would relate to the planarian. Anterior-posterior positional information is represented as a gradient in which different threshold levels along the gradient establish boundaries, dividing the organism into regions labeled A–E. (After Wolpert, 1978, and Gilbert, 1991.)

levels in these regions. Does this in fact occur? Such an experiment would be a good test of the model.

Polar coordinate model

The **polar coordinate model**, proposed by French, Bryant, and Bryant (1976), was first developed to explain patterns of regeneration in vertebrate and insect limbs. It attempts to explain patterns of regeneration in three dimensions. It is a model well worth understanding, and in your studies you can test whether or not it is applicable to planarian regeneration.

The polar coordinate model states that cells recognize their position in the whole because of positional information that each cell receives (like a zip code) designating where the cell is in relation to the rest of the organism. Only two systems of information are needed to generate this zip code: one that gives positional information along the **anterior-posterior axis**, and one that indicates position within the **circumference**. This information is thought to be coded not in diffusible morphogens, but in the molecules found on the surfaces of cells.

The model designates two sets of values that give positional information. The set of values designating position along the anterior-posterior axis is represented by the letters A–E (see Figure 13.3A; the number of letters is arbitrary); the other set of values, designating position in the circumference, is represented as numbers on a clockface (Figure 13.3B; again, the number is arbitrary, but having 12 creates a clever metaphor). The zip code of a cell is its own combination of a letter and number. The model postulates that each cell knows which values it should be sandwiched between. As long as a cell sees the appropriate values surrounding it, it remains happy. If those values are inappropriate, however, regeneration will occur until a correct sequence is reestablished. For example, if a section corresponding to B were removed (Figure 13.3C), allowing sections A and C to be juxtaposed, regeneration would take place and would intercalate (insert) a new section B between A and C. If a slice were taken out of the circumference that, in the clockface metaphor, brought numbers 2 and 5 into juxtaposition (Figure 13.3D), regeneration would occur to intercalate the missing values 3 and 4 between the 2 and 5.

Figure 13.3
Schematic diagram of the polar coordinate model of positional information as it would relate to the planarian. The diagram shows the model proposed by French, Bryant, and Bryant (1976) for vertebrate and insect limbs, as it would relate to the planarian. (A) Anterior-posterior positional information is represented as a series of values ranging from A to E. (B) Circumferential positional information is represented as points on a clockface. Each number on the clockface represents a different positional value. (C) To test the model for anterior-posterior positional information, section B can be removed and the remaining pieces annealed. The model predicts that regeneration will occur to intercalate a new section B between sections A and C. (D) To test the model for circumferential positional information, a pie-shaped slice containing positions 3 and 4 can be removed. Wound healing brings positions 2 and 5 into juxtaposition. The model predicts that regeneration will occur to intercalate positions 3 and 4 between positions 2 and 5.

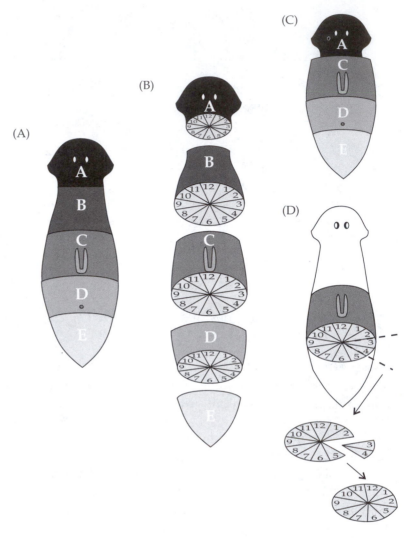

Examples of regeneration in vertebrate and insect limbs have shown that intercalation does occur in these systems, and various manipulations have resulted in often bizarre patterns of regeneration. To explain these patterns two rules of intercalation have been added to the model.

Rule one Intercalation will always proceed by the shortest route to restore positional continuity. If the numbers 2 and 5 are juxtaposed, for example, 3 and 4 will be intercalated rather than 1, 12, 11, 10, 9, 8, 7, 6, 5, and 4—which also would restore continuity but would be the long way around.

Rule two Regeneration will proceed by restoring proximal structures first, and then progressively more distal structures—that is, it will first complete the circle of positional values at the cut surface and then regenerate more distal structures.

Designing experiments to test the polar coordinate model of positional information can be great fun. For example, you can remove a piece of the planarian, rotate the piece, and then replace it (see Figure 13.5A). This provides a myriad of possibilities for confusing polarities. If you rotate it such that the ventral side is now dorsal, you will have inverted the "clockface." The cells of the graft now have circumferential positional values that don't match those of its graft site. The model predicts that there should be regeneration along the edges of the graft, even though technically you haven't removed any tissue from the animal. If the model is correct, you should get duplication of structures in the region of the graft. You can convince yourself of this by diagramming the regenerate, showing clockface values for the graft and the area where the graft is placed. Then, using rule one above, insert the series of numbers that would restore positional continuity. Imagine what this would represent in terms of actual body parts.

Patterns of regeneration

A number of bizarre patterns of regeneration emerge when students are given free reign to cut as they wish. Some of these oddities can be explained very well using the models presented. For example, if a T-cut is made so that the head is removed and the stump is cut longitudinally part way down its length (see Figure 13.4C), two heads regenerate. The polar coordinate model predicts that this would occur, since, on each side of the cut, the region would sense the loss of half its "clock-face," and each half would intercalate the missing values until two complete heads had been formed.

Some of the patterns of regeneration, however, are difficult to explain. If, for example, a short slice is removed from the planarian, the slice may not regenerate into a complete worm. If it is taken from the anterior region and the slice is wider than it is long, it can regenerate a head anteriorly and another head posteriorly (Figure 13.4B). Such a two-headed regenerate has been called "Janus-heads" (an allusion to the Roman god Janus, who guarded gateways by having two faces, one looking forward and another looking back). A narrow slice taken from the posterior region can regenerate two tails, referred to as "Janus-tails" (Figure 13.4B). Can one of the models presented explain these patterns of regeneration? A typical explanation using the diffusion gradient model has been that if a slice is too narrow, the difference in concentration of the morphogen at the anterior and posterior surfaces is too small to be detectable, so the polarity of the piece cannot be established. Do you believe this? Can you come up with a different or better explanation?

Mead and Krump (1986) have done some important work establishing that it is the ratio of length to width in a slice taken from a planarian that determines whether it will regenerate normally, and not the absolute width of the slice. As the ratio of length to width drops below 1.0, abnormal regenerates increase; at a ratio of 0.8, as many as 90% of the regenerates are abnormal. How might you fit this information into one of the models discussed? Can you use it to suggest a different model?

Further experiments only thicken the plot. If a narrow slice is removed, but there is a delay between the time that the first and second cuts are made (see Figure 13.5C), different patterns of regeneration result. If the second cut is made after a delay of two or more hours, regeneration can be completely normal (Child, reviewed in Brøndsted, 1969). If the delay is short, only 5–12 minutes, regeneration is abnormal, and the percentage of abnormal regenerates is actually higher than when the second cut is made almost immediately (in less than 1.5 minutes after the first) (Mead, 1991). Time obviously is an important parameter in establishing positional information. Can you now modify one of our positional information models to explain this, or devise your own explanation? Your ideas will be as good as anyone's.

It is frustrating that we still do not know what the actual substances are that bestow positional information. Several suggestions have been made for anterior-

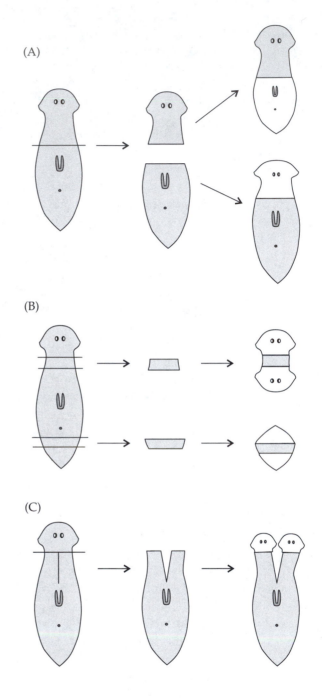

(A)

(B)

Figure 13.4
Patterns of regeneration in the planarian. (A) When the planarian is cut in half, the anterior half regenerates a tail end, and the posterior half regenerates a head end. (B) A thin slice taken from the anterior region can regenerate two heads, a pattern called "Janus-heads." A thin slice taken from the posterior region can regenerate two tails, called "Janus-tails." (C) When a T-cut is made where the head is removed and the stump is split part way down its length, two heads regenerate. Shaded areas represent the original piece; unshaded areas represent regenerated portions.

(C)

posterior gradients that might establish positional information (metabolic gradients; inhibitors and inducers secreted by the nervous system; electric fields; and vitamin A, which acts as a morphogen in some other systems). You could test for these in your experiments. One exciting development is the discovery of homeobox genes in planarians (Garcia-Fernández, Baguñá, and Saló, 1991). This family of genes is known to be involved in pattern formation and appears to be doing the same in the planarian. Two of the planarian homeobox genes (*Plox-4* and *Plox-5*) have been shown to be expressed in an increasing gradient along the anterior-posterior axis of both the intact animal and the regenerate blastema (Orii et al., 1999).

The presence of these homeobox genes in planarians gives us products to be looking for, and suggests genetic manipulations that might crack open some answers.

Experimental Procedures

Planning your experiments

A major task will be to think of clever experiments that test the diffusion gradient or polar coordinate models, or models of your own. A few suggestions are diagrammed in Figure 13.5; they represent only a small part of the smorgasbord of possibilities. You can come up with many, more inventive ones on your own. All of your work will be important to understanding positional information.

You should design your experiments in advance. These should be written in your laboratory notebook, with each experimental plan on a separate sheet, since you will be adding your observations for each experiment to these sheets. You should have a standard format for each experiment in which you:

1. On an outline diagram of a planarian, show exactly what manipulations you plan to do.
2. State what model you are testing and in what ways your experiment tests this model.
3. Describe, with diagrams, what results the model predicts.

Procedures

You have already observed that your animals are energetically motile, and this will be a disadvantage when you start to make your cuts. The animal can be slowed significantly by cooling it. One method is to transfer the animal to a glass slide that is sitting on ice. Since you will have to do all your cuts under a dissecting microscope, realize that the light will be heating up your specimen, counteracting the effects of the ice, so keep the light at a distance if possible. Cuts can be made with microknives or two microneedles pulled across one another in a scissor action. If you use a single microknife to cut, you will find it convenient to have a needle or probe in the other hand to steady the animal during the cutting. A thin slice of a specified width can be made by using two razor blades. Tape the two blades together with a spacer such as a small piece of cardboard positioned between the two blades to give the desired width for the cut. This device also makes the two cuts simultaneously rather than sequentially, which is an important element in the experiment. A hole of a specified diameter can be punched out of the organism by using a Pasteur pipette. Break off the bill of the pipette at the level that will give the bore size you want. The unpolished end of the pipette is an excellent cutting device and can be used as you would use a cork borer.

When the animal is stretched out, quickly make your cut. If you need to manipulate or transfer pieces, use microneedles or a soft paintbrush. You may find that the cuts and manipulations you make are not the ones you had initially planned. Don't despair. Simply change your experimental design accordingly.

If you are doing a grafting experiment, the trick is to align the pieces accurately and then to keep the regenerate relatively still during wound healing so that the graft anneals in place. You undoubtedly will invent your own solutions, but here are some tips. Pieces can be aligned on a piece of wet filter paper kept moist from

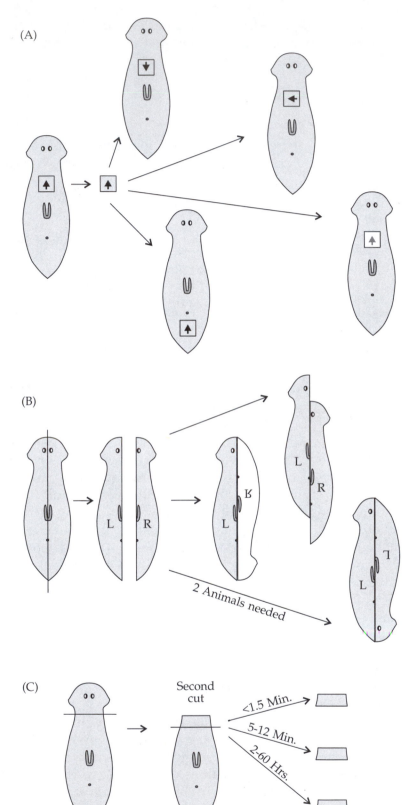

Figure 13.5
Suggestions of experiments that will test models of regeneration. (A) A piece removed from the planarian and replaced in different orientations will juxtapose positional values that don't match. A number of positions can be tried, including flipping the graft so that its ventral side is aligned with the dorsal side of the animal. (B) Different halves of animals can be aligned so that anterior-posterior and circumferential axes are manipulated either simultaneously or independently of one another. (C) The parameter of time can be investigated by altering the time at which a second cut is made when removing a slice from the animal. Shaded areas represent dorsal surfaces of the organism, unshaded areas are ventral surfaces.

below by cotton soaked in ice-cold spring water. In general, a graft that is placed in a hole constructed for it will stay in place more readily than two pieces placed side-by-side. Also, keeping the regenerate cold during wound healing is important for keeping the pieces aligned: a cold room set at 10°C is ideal for this, or you can put the regenerate in the refrigerator for up to 24 hours after the operation.

Whatever cuts you make, record their exact positions using diagrams, and save all your pieces. Each piece can give you information about regeneration.

You will be keeping your regenerates over a period of time. Since regenerates move and climb over partitions in a dish, each regenerate should be put into a separate small petri dish. Be sure to label the dishes with your name and their experiment number.

The water in your cultures must be changed frequently, every other day or so. This is a nuisance, since it's difficult not to throw out the regenerate with the bathwater. Using a pipette to remove the old water is the safest method. Also, remember that chlorinated tap water is toxic. Use only well water or spring water for your cultures.

Record keeping and analysis

Each time you change the water you should record the progress of your regenerates. As you observe them, recall that if the regenerates are kept at room temperature, you can expect to see a blastema within a day or so, differentiated tissues in 4–6 days, and a complete, proportional regenerate in 2–3 weeks. Sometimes it is difficult to determine what you are seeing. Adjust your light for transmitted and then reflected light. Different structures may reveal themselves under different lighting. Also, remember that the blastema will be unpigmented, and that the presence of pigmented eye cups is a sure sign of head regeneration.

Keep careful records of the progress of your regenerates, using diagrams to illustrate what you observe. These should be drawn to scale, since size is data. Record approximately how many times larger you have drawn each regenerate. Make a checklist of behavioral responses that you are monitoring, such as responses to touch and light; speed and type of movement; direction of movement; and the ability of a regenerate to right itself. Test for these each time you make a set of observations. Compare these responses to that of your control animal.

Do not feed cultures during regeneration. You are not being cruel; they will not be capable of feeding until regeneration is complete, and food in the water will simply rot and contaminate your cultures. If any of your tissues die during the experiment, remove them so that they don't become a source of contamination.

When you start analyzing your data, present it in chart and graph form wherever possible. This will simplify making comparisons among different experiments and will allow you to see trends and the timing of events more easily. Realize that in the best of all possible worlds you would have repeated experiments a number of times and would have been able to report the frequency with which a certain pattern of regeneration occurred. (Frequency is important data.)

Determine whether your results are similar to those predicted by the model you are testing. If they are different, try to reformulate the model so that it conforms to your results or create your own model. This is no simple chore, but it is true science. In your excitement and frustration, keep in mind that it is this method of investigation—designing models and then testing them—that fuels advancing science at explosive speeds. Your fuel may be the high-octane we've been looking for.

Accompanying Materials

Tyler, M. S. and R. N. Kozlowski. 2000. *Vade Mecum: An Interactive Guide to Developmental Biology*. Sinauer Associates, Sunderland, MA. "Flatworm." This chapter of the CD illustrates the anatomy and behavior of the flatworm as well as various regeneration patterns in labeled photographs and movies.

Gilbert, S. F. 2000. *Developmental Biology*, 6th Ed. Sinauer Associates, Sunderland, MA. Chapters 3 and 18. These pages give concise explanations of diffusion gradient models and the polar coordinate model as they relate to regeneration.

Selected Bibliography

Baguñá, J. 1998. Planarians. In *Cellular and Molecular Basis of Regeneration: From Invertebrates to Humans*, P. Ferretti and J. Géraudie (eds.). John Wiley and Sons, Ltd., New York, pp. 135–165. A superb and comprehensive review of what is presently known about planarian regeneration.

Baguñá, J., E. Saló, J. Collett, C. Auladell and M. Ribas. 1988. Cellular, molecular, and genetic approaches to regeneration and pattern formation in planarians. *Fortschr. Zool.* 36: 65–78. A concise, well-written review of planarian regeneration studies, dealing less with theoretical models than with descriptive analysis of the process.

Brøndsted, H. V. 1969. *Planarian Regeneration*. Pergamon Press, Oxford. This is a classic. Its almost encyclopedic record of previous experiments is invaluable. If in riffling through its pages you find the same experiments you are doing, do not be discouraged. All good experiments are worth repeating.

Chandebois, R. 1984. Intercalary regeneration and level interactions in the freshwater planarian, *Dugesia lugubris*. II. Evidence for independent antero-posterior and medio-lateral self-regulating systems. *Roux Arch. Dev. Biol.* 194: 390–396. This is the second in a series of three papers in the same journal. The inventiveness of the experiments to test models of regeneration make the papers well worth examining.

French, V., P. J. Bryant and S. V. Bryant. 1976. Pattern regulation in epimorphic fields. *Science* 193: 969–981. The authors of the polar coordinate model excited the world of regeneration studies with this presentation of their model. Though very readable, it can be found in an even more palatable style in a later article, "Biological regeneration and pattern formation," in the July 1977 issue of *Scientific American*.

Galtsoff, P. S., F. E. Lutz, P. S. Welch and J. G. Needham (eds.). 1959. *Culture Methods for Invertebrate Animals*. Dover Publishing, New York. This is a republication of a work originally published in 1937 and is still worthy of another printing. It is an excellent source of the nitty-gritty details that are needed for culturing animals in the laboratory.

Garcia-Fernández, J., J. Baguñá and E. Saló. 1991. Planarian homeobox genes: Cloning, sequence analysis, and expression. *Proc. Natl. Acad. Sci. USA* 88: 7338–7342. A truly exciting paper, this marks the first genetic study to find homeobox genes in the planaria.

Gilbert, S. F. 1991. *Developmental Biology,* 3rd Ed. Sinauer Associates, Sunderland, MA. An earlier edition of the text currently in use. It includes a diagram on page 615 that is useful in studying the gradient diffusion model of positional information.

Goss, R. J. 1969. *Principles of Regeneration.* Academic Press, New York. This is a delightfully written book. It is ageless and well worth reading with an excellent chapter on planarians.

Gremigni, V. 1988. Planarian regeneration: An overview of some cellular mechanisms. *Zool. Sci.* 5: 1153–1163. A concise review, this concentrates primarily on blastema formation and the origin of blastema cells.

Kato, K., H. Orii, K. Watanabe and K. Agata. 1999. The role of dorso-ventral interaction in the onset of planarian regneration. *Development* 126: 1031–1040. An intriguing paper that suggests that the trigger to regeneration in planarians is the juxtaposition of dorsal and ventral epithelia during wound healing.

Malacinski, G. M. and S. V. Bryant (eds.). 1984. *Pattern Formation: A Primer in Developmental Biology.* Macmillan, New York. An important collection of papers on regeneration. Though none is specifically on planarian regeneration, a number of models are discussed, including the polar coordinate model.

Mead, R. W. 1991. Effect of timing of cutting on patterning and proportion regulation during regeneration of the planarian *Dugesia tigrina.* In *Turbellarian Biology,* S. Tyler (ed.). Kluwer Academic Publishers, Dordrecht, Netherlands, pp. 25–30. This is one of the landmark papers in planarian regeneration that will help to fine-tune investigations. By showing that small increments of time are important factors in pattern formation, it will allow geneticists and biochemists to focus their studies within specific time frames.

Mead, R. W. and J. Christman. 1998. Proportional regulation in the planarian *Dugesia tigrina* following regeneration of structures. *Hydrobiologia* 383: 105–109. A study that looks at the "normalizing' of proportion in regenerates after structures have reformed.

Mead, R. W. and M. A. Krump. 1986. Abnormal regeneration in the planarian *Dugesia tigrina* as a function of the length:width ratio of the regenerating fragment. *J. Exp. Zool.* 239: 355–364. Another paper from Robert Mead's laboratory; this one focuses attention on the fine parameters of regeneration, showing that the ratio of dimensions of a piece of planarian rather than absolute dimensions determines patterns of regeneration. This is an extremely important clue for figuring out pattern formation and must be fit into any model of positional information.

Morgan, T. H. 1898. Experimental studies of the regeneration in *Planaria maculata. Arch. Entwicklungsmech. Org. (Wilhelm Roux)* 7: 364–397. For historical reasons alone this should be fascinating to the student of regeneration. The first paper by Thomas Hunt Morgan on his planarian research, it catalogues a number of his experiments and shows the style in which science was reported in these early days.

Orii, H., K. Kato, Y. Umesono, T. Sakurai, K. Agata and K. Watanabe. 1999. The planarian HOM/HOX homeoboxgenes (*Plox*) expressed along the anteroposterior axis. *Dev. Biol.* 210: 456–468. This excellent paper gives evidence that homeobox genes help establish anterior-posterior patterning in planarians. It includes a useful chart comparing homeobox genes of a number of organisms.

Rieger, R. M., S. Tyler, J. P. S. Smith III and G.R. Rieger. 1991. Platyhelminthes: Turbellaria. In *Microscopic Anatomy of Invertebrates, Vol. 3: Platyhelminthes and Nemertinea,* F. W. Harrison and B. J. Bogitsch (eds.). Wiley-Liss, New York, pp. 7–140. This is an impressive, well-written chapter on turbellarian biology. You will not find anywhere else such a complete, up-to-date review of all that is known at the light-microscopic and ultrastructural levels about the anatomy of these animals.

Winston, J. E. (ed.). 1999. Libbie Henrietta Hyman: Life and Contributions. *American Museum Novitates,* Number 3277. American Museum of Natural History, New York. This entire issue of *Novitates* is devoted to articles on Libbie Hyman, one of the giants in the field of invertebrate zoology who devoted much of her life to the study of flatworms.

Wolff, E. 1961. Recent researches on the regeneration of planaria. In *Regeneration,* D. Rudnick (ed.). Ronald Press, New York, pp. 53–84. This is an important chapter summarizing the model proposed by Etienne Wolff that gradients of inducer and inhibitor substances are the basis of positional information in planarians.

Wolpert, L. 1978. Pattern formation in biological development. *Sci. Am.* 239(4): 154–164. This is a beautifully illustrated article, written by the author of the term, "positional information." It provides an excellent, clear presentation of various models of regeneration.

Suppliers

Dugesia dorotocephala **and** *D. tigrina* **can be obtained from:**

Connecticut Valley Biological Supply Co., Inc.
P.O. Box 326
82 Valley Road
Southampton, MA 01073
1-800-628-7748

Carolina Biological Supply Co.
2700 York Road
Burlington, NC 27215
1-800-334-5551
www.carosci.com

chapter *14* *Amphibian Development*

Breeding

In areas where snow blankets the ground for much of the winter, one of the surest signs of spring is a chorus of frogs cracking the stillness of evening with their din or, more subtly, the appearance of spermatophores left on the leafy bottom of a temporary pool by an amorous salamander. It is the male amphibians that respond first to the new pulse of spring. Male frogs and toads (anurans, meaning "no tail") make all the racket at the breeding pools belting out their "song." Male salamanders (**urodeles**, meaning "tailed"), lacking vocal cords, quietly lure their mates with a complex and intimate dance.

The females arrive at the breeding pools more sedately. Attracted by the male's song or dance, they choose their mate. Mating differs according to the mode of fertilization. Fertilization in most anurans is external, and to ensure a successful mating, the sperm must be deposited on the eggs as they are laid. Male frogs and toads, therefore, usually grasp the female around the abdomen in an embrace called **amplexus**. This stimulates the female to extrude her eggs, and as she does, the male releases his sperm over the eggs. This usually is the last the parents concern themselves with their offspring. After mating, most frogs and toads abandon their progeny to the elements.

Most salamanders have internal fertilization. For mating to take place, a female must pick up a packet of sperm, the **spermatophore**, which the male lays on the bottom of a pool. The female chooses her partner as he woos her with his courting dance, a series of weaves and caresses; the dance brings the female over the spermatophore, which she clasps with the lips of her cloaca. Once lodged in the cloaca, the tip of the spermatophore releases its sperm, and the sperm find their way to a specialized region of the cloaca called the **spermatheca**. Here they are stored until the female starts laying her eggs. As the eggs pass by the spermatheca, they are fertilized by the sperm. A female may lay her eggs soon after mating, or she may wait months, depending on the species. Usually the site for laying is a shallow pool or a well protected, moist nest. In those species that have nests, the female often guards her eggs until hatching.

Some amphibians return to the same breeding site year after year, and certain species of salamanders are known to return exclusively to the site where they were raised as larvae. This homing behavior makes these amphibians especially sensitive to the destruction of their habitat—a problem that today threatens so many species of plants and animals.

Environmental Hazards Affecting Amphibian Development

For decades, a local pond or woods has provided the developmental biologist with a richly stocked field laboratory for amphibian studies. In recent years, however, this laboratory has become poorly stocked in many regions. Declines in amphibian populations have been reported over wide geographic areas: in the Americas, Britain, Australia, and the Amazon basin. In the United States, extensive declines have been documented for the Northwest and Sierra Mountains; they also are reported for the Northeast and Southeast. The decline appears to be due primarily to the loss of wetlands, increased ultraviolet radiation, and the pollution of the breeding sites that remain. Amphibians are especially vulnerable to these habitat changes during their developmental stages, because in most species the embryos and larvae develop in water, and the eggs and embryos, being surrounded only by permeable jelly coats and a vitelline membrane, are unprotected from pollutants.

The individual environmental factors that can adversely affect amphibian development are numerous and together create a complicated interplay of effects. One of the culprits is acidic precipitation. Acid rain and snow have lowered the pH of breeding sites, affecting the pH levels of temporary pools more significantly than those of permanent ponds. It is estimated that roughly 10–27% of temporary ponds in the Northeast have pH levels lower than 5.0. Since temporary pools provide the breeding sites of roughly 30% of all salamander species and 50% of all frog species in North America, their acidity could have a significant effect on populations. Sensitivity to acidity varies widely among different species. Though the embryonic stages of some species show complete mortality at pH levels of 4.0–5.0—a typical range for an acidified pool—others survive well in this range.

Acidity is not the only culprit in this story, however; it acts in concert with other factors. Lowered pH levels cause aluminum to be leached out of the substrate, and high levels of aluminum have been found to be very toxic to the early larval stages of amphibians. But the formula for toxicity is not simple. The toxicity of the aluminum is very dependent on the pH of the water, and the combination of aluminum and pH that is lethal to one group can be quite different from the combination that affects another group.

Another severe threat to amphibian development is long stretches of dry weather that dry up temporary pools. This quickly kills any embryos and larvae that are left in the pools. Acid rain, once again, can be a contributor to the problem. Acidity that may not be low enough to kill the embryos can still slow larval growth. The extended period required for these larvae to come to metamorphosis becomes critical if it leaves the larvae at the bottom of a dry pool.

Sudden temperature changes can also be a threat to the survival of amphibian embryos. Though the embryos usually have a wide range of temperature tolerance (the range for *Rana sylvatica* embryos, for example, is 5–21°C), a sudden freeze can kill off any embryos that are caught near the water's surface. More problematic can be the synergistic effects of temperature and acid stress. An embryo developing at the top of its temperature range will succumb if exposed to an acid stress as well.

Another hazard to amphibian development, one that has received attention lately with the thinning of the protective upper atmospheric ozone layer, is the increased exposure to ultraviolet radiation. This can be particularly damaging to amphibians because of the long exposure of their embryos to sunlight. In addition, the synergistic effects of ultraviolet radiation and low pH are particularly damaging and can greatly increase mortality.

You will be measuring only a few of the environmental factors that could be affecting amphibian development in your local areas, since other factors, such as aluminum levels and synergistic effects, are probably beyond your monitoring abilities. However, by monitoring pH, temperature, and the drying out of pools, and correlating this with survival rates of the embryos you observe, you should get a general picture of amphibian reproductive success in your area.

Preparing for the Field Trip

Learn about your local amphibians first

Most amphibians in the Northeast breed in the spring or early summer. Those in the Southeast may breed even in the winter months, coming out in a warm rain. In the more arid West, breeding is often rhymed closely with periods of rain. The eggs—whether hidden under rocks or logs, attached snugly to submerged vegetation, or broadcast haphazardly to float in rafts at the water's surface—are usually easily found if you know when and where to look. So schedule your trip for a time when several species of amphibians are breeding (this will differ radically, depending on your geographic region. Check a regional guide to amphibians that gives you specifics about your area; some are listed in the bibliography.) Table 14.1 is a short listing that will help. Determine which species you can expect to find in your area and where you might find them. It is always advisable to consult a herpetology enthusiast about their field notes on "who, where, and when." (These invaluable people quietly exist everywhere—just start asking around. Often they will be willing to come along on your field trip.) Determine a route for your trip that will pass through both woods and fields and will encounter a pond, temporary pools such as ditches, and slow-moving water. Swamps and marshes are wonderful if you can find them. Try to schedule both a daytime and a nighttime field trip. Eggs are most easily found by day, but the night brings forth both mating behavior in salamanders and frogs and the varied choruses of the frogs.

Equipment needed and recording of data

For the field trip, you will need long-handled fish nets for egg masses beyond arm's reach; clean plastic (because glass breaks) containers with tops; index cards; a pencil (because pen runs); a measuring tape; a sturdy thermometer (a metal pocket thermometer is especially nice); some distilled water (for washing electrodes if necessary and wetting soil samples); and a pH-measuring device. (There are all levels of sophistication of pH meters, and they vary in reliability; see the section on pH measurements. Choose a method that fits your goals and purse.)

Whenever you find a clutch of eggs, record information about the eggs on an index card, and place the card along with a container of three or four eggs

Table 14.1 Breeding information on some commons North American amphibians

Species	Distribution	Season eggs laid	Laying habitat	Where eggs are found	Egg clutch size	Egg
Frogs						
Acris gryllus Cricket frog	Southern and central U.S.	February–October	Shallow meadow pools and swamps	Attached to stems of grass in shallow water or strewn on bottom	Eggs laid separately, few up to 250, jelly firm	0.9–1.0 mm diam. Brownish above, white below
Hyla regilla Pacific tree frog	Western U.S. and Canada	January–May	Quiet water, pools, ponds, swamps	Attached to vegetation at or below water surface	Eggs laid in small, loose, mass, 10–70 eggs, loose jelly	1.3 mm diam. Brown above, yellowish below
Hyla versicolor Gray tree frog	Eastern U.S. and southeastern Canada	May–June	Temporary pools or quiet permanent water	Loosely attached to vegetation at water surface	10–40 eggs, loose jelly	1.1–1.2 mm diam. Very light brown to gray above, white to yellow below
Pseudacris (Hyla) crucifer Spring peeper	Eastern U.S. and southeastern Canada	April–May	Temporary or permanent water	Eggs attached singly to submerged vegetation	Up to 900 eggs, jelly firm	1.1 mm diam. Dark brown-black above, white below
Rana catesbeiana Bullfrog	East of Rockies and West Coast	May–August	Permanent water with vegetation	Thin, floating mass at surface	12,000–20,000 eggs in mass up to 0.6 m across, jelly loose	1.23–1.7 mm diam. Black above, creamy white below
Rana clamitans Green frog	Eastern North America	June–August	Permanent quiet water with vegetation	In shallow water, floating mass, attached to vegetation	1500–5000 eggs in mass 15–30 cm diam., jelly loose	1.5 mm diam. Black above, white below

Species	Range	Breeding season	Habitat	Egg deposition	Egg mass	Egg description
Rana palustris Pickerel frog	Eastern North America	April–May	Permanent water with vegetation	In center of water, attached to submerged sticks	2000–3000 eggs in globular mass, 5–10 cm diam. Firm jelly	1.6 diam. Brown above, white to yellow below
Rana pipiens (species complex) Leopard frog	U.S., Canada, and Mexico, except for West Coast	March–May (after wood frog) in North. February–December in South	Permanent water with vegetation such as lake inlets, slow streams, ponds, or overflows	Laid in the shallows, often in communal masses	2000–4000 eggs in flattened oval masses, 7.5–15 cm × 5–7.5 cm, jelly loose	1.6 mm diam. Black above, white below
Rana sylvatica Wood frog	Northern U.S. and Canada	March–May	Temporary pools	Often in large communal masses, attached to submerged stems or branches	Up to 3000 eggs in globular mass, 6–10 cm diam. Jelly firm	2.0 mm diam. Black or chocolate above, ivory white below
Toads						
Bufo americanus American toad	Eastern and Midwest U.S. north of Georgia, southeastern Canada and west through Ontario	April–May	Shallow, open pools	Threaded among submerged vegetation or stretched out on bottom	6000 eggs in pairs of long (4 m or longer), spiral strings, eggs in single row down string, jelly firm	1.0–1.4 mm diam. Black above, white below
Bufo boreas Western toad	Northwestern U.S. and Canada	March–September	Ponds and shallow pools	Long strings strewn among grasses usually at margins of pool or in shallow water	Up to 17,000 eggs in long strings, eggs in one or two rows down string; jelly firm	1.5–1.75 mm diam. Black above, white below
Bufo cognatus Great plains toad	Dakotas south through Texas and Southwest to Calif.	April–September	Pools, irrigation ditches, overflow from streams	Long strings in shallow water	Up to 20,000 eggs in long strings, eggs in one or two rows down string, jelly firm	1.18 mm diam. Black above, white below

(continued)

Table 14.1 Breeding information on some common North American amphibians *(continued)*

Species	Distribution	Season eggs laid	Laying habitat	Where eggs are found	Egg clutch size	Egg
Bufo woodhousii fowleri Fowler's toad	Central and eastern U.S.	April–June	Shallow water of pools, lake margins, or ditches	Laid in strings tangled in aquatic vegetation below surface	8000 eggs, in long tangled strings, eggs in one or two rows down string, jelly firm	1.0–1.2 mm diam. Black above, white below
Bufo woodhousii woodhousii Woodhouse's toad	Great Plains and western U.S. to Calif.	March–July	Shallow water of pools, lake margins, or ditches	Laid in strings tangled in aquatic vegetation below surface	Up to 25,000 eggs, in long tangled strings, eggs in one or two rows down sting, jelly firm	1.0–1.2 mm diam. Black above, white below
Salamanders						
Ambystoma jeffersonianum Jefferson salamander	Eastern U.S. and southern Canada	Early Spring	Temporary or permanent water	Laid singly, attached to one another and to a twig.	100–300 eggs, jelly capsules often green from unicellular algae, jelly thick and soft	2–2.5 mm diam. Dark brown-black above, lighter below
A. laterale Blue-spotted salamander	Northeastern U.S. and Canada	April	Shallow (30–40 cm deep or less) temporary pools with vegetation or leafy bottoms	Laid in loose groups that attach lightly to vegetation or fall to bottom	Single or loose groups of 2–6 eggs, large clear jelly capsules	2–2.5 mm diam. Dark brown-black above, lighter below
Ambystoma maculatum Spotted salamander	Eastern U.S.	January–May	Temporary pools	Attached to sticks about 15 cm below surface	100–200 eggs in oval mass about 10 cm across, jelly firm, often milky opaque	2.5–3 mm diam. Dark brown or gray above, whitish to yellow below

Species	Range	Season	Location	Egg placement	Egg mass	Size/color
A. opacum Marble salamander	East and Midwest, not north of Boston–Chicago line	September–October	Under moss or log at edge of water or in dry beds of temporary ponds or streams	Laid singly in a depression likely to be flooded by fall rains	50–232 eggs, guarded by female, two separate jelly envelopes easily distinguished	2.7 mm diam. Dull yellow
A. tigrinum Tiger salamander	U.S, Canada, Mexico	December–April	Temporary pools	Attached to twigs about 1 inch below surface	23–110 eggs, globular or oblong mass, 55–70 mm ×75–100 mm, jelly loose	2.5 mm diam. Light to dark brown above, pale cream to buff below
Eurycea bislineata Two-lined salamander	Central and eastern U.S.	April–June	Flowing water of small brook	Cemented individually by stalk to underside of rocks	12–36 eggs, two distinct jelly envelopes, outer one extended into slender jelly stalk	2.5–3 mm diam. White to pale yellow
Hemidactylium scutatum Four-toed salamander	Eastern U.S.	Mid-April–June	Cavities, 7–10 cm deep, in moss or grass overhanging water	Laid singly but adhere in cluster	19–50 eggs, not aggregated into common jelly envelope, but cling to one another, not stalked, guarded by female	2.5–3 mm diam. Very light pigment above, yellowish below
Necturus maculosus Mudpuppy	East of Rockies, north of Georgia	Spring and summer	Lakes or streams	Excavations scooped out from underneath large rocks or submerged logs	60–100 eggs, attached singly by stalks to roof of nest, three jelly envelopes, guarded by female	5–6 mm diam. Light yellow
Notophthalmus viridescens Eastern newt or Red-spotted newt	Eastern U.S, southeastern Canada	April–June	Sluggish water	Attached singly to submerged stems, leaves or wrapped individually in leaves	20–30 eggs, separate, not in common mass, elliptical jelly envelopes, outer jelly covered with thin filaments	1.5 mm diam. Light to dark brown above, yellowish green below

(continued)

Table 14.1 Breeding information on some common North American amphibians *(continued)*

Species	Distribution	Season eggs laid	Laying habitat	Where eggs are found	Egg clutch size	Egg
Plethodon cinereus Red-backed salamander	Northeastern U.S.	June–July	Cavities in well-decayed logs or stumps	Suspended from roof of nest in small clusters	4–17 eggs, outer jelly envelope covered with white mucous strands, converge to form attachment stalk, held together like bunch of grapes and suspended by common pedicel, guarded by females	3.0–5.0 mm diam. White above, very pale yellow to pale yellow-orange below
Taricha granulosus (*Triturus granulosus*) Rough skin (Oregon) Newt	West Coast, S.E. Alaska-central Calif.	Late December–July	Streams, lakes, and ponds	Laid singly or small clusters, attached to vegetation and other objects in water	Single or 2–3 eggs, outer jelly envelope 3.63 mm, jelly firm	1.8–2 mm diam. Light brown above, white below
Taricha torosus (*Triturus torosus*) California newt	California coastal ranges	December–May	Streams, ponds, and reservoirs	Laid in small clumps, attached to sticks, vegetation, and other objects under water	7–29 eggs, globular mass, 15 mm across, jelly firm	1.88–2.5 mm diam. Pale brown above, whitish-yellow below

Source: Primarily after Hunter, Albright, and Arbuckle, 1992; Rugh, 1948; and Wright and Wright, 1949.

(more is unwarranted) and water or soil from the site. Your collection of eggs will be brought back to the lab for measurements and observations of development. If there are only very few eggs at the site, you will *not* be collecting any, but record the information about them. The information you take down is important for making reliable identifications of the species of eggs, for updating records on "who, what, where, and when" amphibians breed in your area, and for evaluating whether and which amphibians in your area are under stress from environmental factors.

The information recorded on the card should be as listed below. It is useful to prepare cards in advance that state the information that is needed and blanks for filling in. Working in teams of three or four lightens the load.

Personnel and date   Who is collecting the data, and the date.

Macrohabitat   Where were the eggs found (e.g., bog, marsh, woods, field)? If a woods, type of trees. Record air temperature.

Mesohabitat   Were the eggs found in a lake, stream, pond, temporary pool, under a log, etc.?

Microhabitat   If eggs are found in water, what is its temperature and pH? How deep is it? Is it still or running? How much vegetation? Is the bottom grassy or leafy? What type of leaves? If eggs were found under a log, what is the temperature and pH of the substrate? Is the log rotten? What type of tree did it come from? Are the eggs laid in moss or leaf detritus? What type of leaves?

Where in specific habitat eggs found   If eggs are found in water, are they at the water's edge or in the middle of the pond? Are the eggs floating or submerged? If submerged, how far under the water's surface? Are they attached? If so, to what? If in a nest, what are its dimensions? Is it near any water?

Description of egg mass   Are the eggs laid singly or in a mass? What are the dimensions of the egg mass? How many eggs in the mass? (If there are too many eggs to count, you can estimate by counting the number of eggs in a small sample, calculating from their measurements the volumes of the small sample and the entire mass, and figuring from these numbers the total number of eggs.) Is the jelly firm or loose? Is it milky or clear? Is it green (contains green algae)?

Description of egg   Color, both above and below; diameter; stage of development (see staging series in Table 14.3); percentage of healthy eggs; description of unhealthy eggs.

Species of the egg   From your information above and that given in Table 14.1, what species do the eggs appear to be? You will be verifying your identification back in the laboratory.

Equipment for pH measurements in the field

It is always best to make your pH measurements in the field rather than back in the lab. This will give the most accurate readings, since the pH of water changes as it is trundled around and loses CO_2. The best method for accurate pH measurement is to use a portable electric pH meter. The Orion™ meter with an Orion Ross™ electrode is particularly accurate, but extremely expensive. Far less expensive meters are available (see the Suppliers list at the end of the chapter). Those that both read and adjust for temperature are the best.

If you use an electric pH meter, you should test your meter for its sensitivity to ionic effects. The low ionic concentrations of the soft water found in ponds and pools make it difficult for electrode pH meters to give accurate readings. After calibrating your pH meter with a known acidic pH buffer, measure a low ionic acidic solution, also of known pH, to determine if there is any fault in the reading. For a low ionic acidic solution, you can use a 0.0001 N solution of hydrochloric acid (HCl). This will have a pH of 4.0. If the meter can't read this dilute solution, try another electrode. Be sure the pH buffer and low ionic acidic solutions are at approximately the same temperature. Always rinse the electrode several times with distilled water between buffers and before reading each sample, and make two or three readings of any sample to ensure accuracy.

Colorimetric pH measurement kits also exist, and these can be relatively inexpensive and convenient for field work. But a word of caution about their use. They have been found to give inaccurate readings in acidic soft water. Discrepancies are usually in the range of ±0.2 units of pH, but can be as high as 0.8 pH units. These kits use dyes of different pH that change color as the pH changes. The color change is measured against a color chart to determine the pH of the solution. You will get the most accurate readings by using dyes that have a pH that is at or very near to the pH of the solution being measured.

Paper pH test strips are the least expensive method of measuring pH, and those with the range most appropriate for this study are sold as acid rain test kits and soil pH strips. They are the least reliable method of measuring pH, but are usually within a lab's budget.

Come dressed for the field

On the day of the field trip, come dressed for the weather, and wear boots (or waders if you have them) and polarized sunglasses. The polarized glass will allow you to see through any glare on the water surface, and you'll spot eggs that would otherwise be invisible. If the field trip is at night, or in the evening extending into dark, bring a flashlight and red cellophane to cover the light. Amphibians are insensitive to red light, so you'll be able to observe their behavior without disturbing it. If it is raining, rejoice. Amphibians love wet weather and will be out in great numbers.

Know government regulations

States and provinces often have regulations regarding amphibians, so check with the appropriate Nongame and/or Endangered Species Program or Department of Wildlife, Natural Resources, or Conservation before embarking on your trip.

The Field Trip

As you trudge along your planned route, examine all bodies of water, regardless of size. Often, it will be the tire rut filled with rainwater that some amphibian has chosen as a nursery for her brood. As a rule of thumb (though there are many exceptions), you can expect to find the eggs of frogs and toads in ponds or temporary pools, and those of salamanders in temporary pools or under and inside logs. At first you may have trouble spotting an egg mass. Wear your polarized sunglasses to reduce glare on the water, get down on your knees, and paw around. Egg masses that are submerged, and especially those sitting on the bottom of a pool or pond, become covered with sediment. Often all that you see to alert you to their presence is a globular shape.

In early spring in the Northeast, you can expect to find the eggs of the wood frog (*Rana sylvatica*); spring peeper (*Pseudacris crucifer* previously *Hyla crucifer*); leopard frog (*Rana pipiens*); and spotted salamander (*Ambystoma maculatum*); all of these are found in water. These eggs are easily distinguishable (some are shown in Figure 14.1). The eggs of the spring peeper will appear singly, at the edges of pools, and are very difficult to see. The eggs of spotted salamanders are in globular masses with firm jelly. The jelly is sometimes milky and opaque. The eggs of the wood frog are slightly smaller than those of the spotted salamander and are usually found in larger masses, with jelly that is less firm. The egg masses of the leopard frog usually are huge and black with loose jelly and are the most easily spotted, but they may be rafted out in the middle of the pond where you can only get to if you brought waders. The communal rafting of these eggs is an adaptation that increases survival, since the center eggs retain more heat and are protected from sudden drops in temperature.

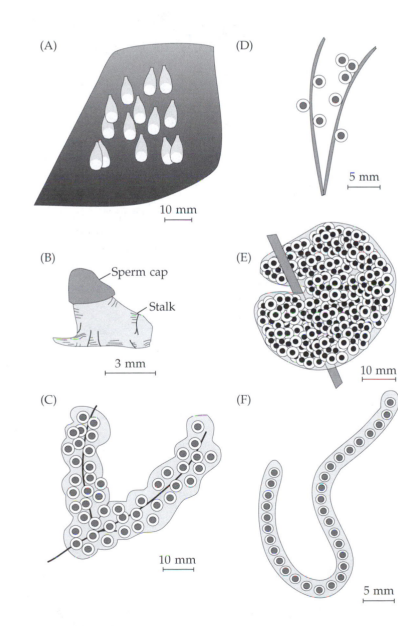

Figure 14.1
Egg masses and a spermatophore of some common North American amphibians. (A) Eggs of the mudpuppy, *Necturus maculosus*, hanging by jelly stalks from the underside of a rock in a lake. (B) A spermatophore of the spotted salamander, *Ambystoma maculatum*. The gelatinous stalk supports a sperm cap. (C) Egg mass of the Jefferson salamander, *Ambystoma jeffersonianum*, laid on a twig in a vernal pool. (D) Eggs of the spring peeper, *Pseudacris crucifer*, laid on grass blades submerged at the edge of a small pond. (E) Eggs of the leopard frog, *Rana pipiens*, laid in a floating mass on the surface of a pond. (F) String of eggs of the American toad, *Bufo americanus*, which would be laid as paired strings on submerged vegetation or on the bottom of a shallow pool.

The normal bicoloration of amphibian eggs is an adaptation as well. The dark upper (**animal**) hemisphere is difficult to see against the dark water when viewed from above and therefore protects against aerial predators. The light lower (**vegetal**) hemisphere is difficult to see for any hungry swimmer looking up at the eggs against a light sky. When such bicoloration is not needed for camouflage, such as in eggs that are laid under rocks or logs, the eggs often are unpigmented.

For any egg mass that is found, record the necessary information on an index card. Remember, working in teams will reduce the amount of work for each of you. Make your pH and temperature measurements on water from the exact location and depth at which the eggs were found. If there are more than 50 eggs in the mass, or there are a number of similar masses in the vicinity, remove three or four eggs and put these in a plastic container along with the index card and enough water to fill half the container. These will be brought back to the laboratory for observation. If you find eggs under a log, these are likely to be those of a salamander, and the female may be around guarding them, so look carefully for her. In general, these eggs should not be disturbed, since usually there are very few. If there are plenty (over 50), two or three may be collected along with a small soil sample. Measure the temperature of the soil and take a pH reading, moistening the soil first with some distilled water.

Remember that we are concerned with possible environmental effects on developmental success, so look carefully at the eggs. It is important to determine how many look healthy and how many appear moribund (abnormally whitish or deteriorating). If there are deformities, what are they, and how many eggs are affected?

Don't forget to look for the spermatophores left by male salamanders (see Figure 14.1). These little cone-shaped, gelatinous structures can be spotted from a moving bicycle, if you know what you're looking for. When you are going at a speed of 15 miles per hour, they often look like the white triangles of hardened sap seen on pine cones (and I've come to a screeching halt many times mistaking the two). They are usually found in clusters of ten or more, often on leaves in the bottom of water-filled ditches bordering woods. If you find some, take one or two back to the lab. If you are lucky, the sperm cap will still be intact, and you can put it on a slide under the microscope to see the living sperm.

During your trip, cover as many habitats as possible, and always observe good ecological manners.

Rules for the Road

Walk softly and carry a long-poled fishing net.
Be inquisitive—turn over rocks and logs, search through wet pond grasses.
Always replace the rock or log and leave the habitat as you found it.
Never tromp as a large group through a pond or other small body of water. Send out a single scout, so as to leave the habitat as undisturbed as possible.
Be patient—pause for long minutes to listen and look.
Never overcollect. Observe and put back.
If you catch an adult amphibian, wet your hands before holding it, and hold it only for a short period. Do not pass it around. Its skin is fragile and dries out very quickly.

Once you have finished your observations on the eggs brought back to the laboratory, *always* return them to their specific habitat as soon as possible.

Night Sounds

If you are lucky enough to schedule a night frog walk, you may be serenaded by a cacophony of songs from males of various species of frogs and toads. Marshes and swamps are among the best places to hear such a concert. If possible, find an anuran enthusiast, ask them which anurans call when in your area, and bring them along on your trip. A very useful aid to prepare you for your walk is a narrated CD or cassette tape of night sounds, which you should listen to several times before your walk. (Two such recordings are *A Guide to Night Sounds* and *The Calls of Frogs and Toads*, listed in the bibliography. They are inexpensive and easily obtainable.) Also, Table 14.2 lists some common North American frogs in the order that they are heard in the spring. The time that they are heard will vary among different regions. You should be able to extrapolate from the table approximate times when you might hear these frogs in your area. Though air and water temperatures are listed for when the males first start calling, this is very approximate, since precipitation is an equally important factor. In general, calling will start after a period of rain. The chart's descriptions of the songs, of course, are subjective, and if you compose better ones, feel free to amend the chart.

Back at the Laboratory

Species identification

Your first job back in the laboratory is to verify your field identification of the species of your eggs. Recheck your identifications against Table 14.1, then use your dissecting microscope to look at each group of eggs you collected. If your eggs are in the early stages of development, you can measure the diameters of the eggs and their surrounding jelly capsules (there should be one or more of these), and use these in your identification. Compare the measurements with those given in Table 14.1 and Figure 14.2. Correct your data card if necessary, and make your collection available for others to see.

Normal development

Now examine your eggs to observe as many stages of development as possible. Notice that the egg itself is not transparent. This is due to the large amount of yolk stored within, concentrated primarily in the vegetal hemisphere. The egg is classified as a **mesolecithal** egg, meaning that it contains a moderate amount of yolk. Most eggs are pigmented with melanin, a black pigment, distributed primarily in the animal hemisphere, giving the egg a bicoloration of dark above and light below. The concentration of melanin varies among the species. The less melanin, the lighter brown the egg. For years embryologists have sought out light-colored amphibian eggs to use for fate-map experiments. The light pigmentation allows for experiments using vital dyes in which a small number of cells on the surface are stained and then followed throughout gastrulation.

Table 14.2 Songs of some common North American male anurans

Species	Time calling is heard in various regions	Air/water temperature when males first start calling	Male mating call
Acris gryllus Cricket frog	Late January–October in North Carolina	/Water can have a surface film of ice	Sounds like a rattle or metallic click in rapid succession
Rana sylvatica Wood frog	Late March or early April-late April, early May in central Maine	50°F/54°F	Short abrupt "wrunk," emitted singly or several in rapid succession, sounds like quacking of ducks
Pseudacris crucifer Spring peeper	Early April–May in central Maine, greatest on warm, rainy nights	52°F/above 50°F	Simple loud peep ending with an upward slur, repeated 15–25 time/minute, occasionally a short trill
Rana pipiens Leopard frog	Late April, early May in central Maine for 1–4 weeks	55°F/59°F	Long drawn out guttural snore up to 3 sec., release call is a short series of trills, and warning is an untrilled scream
Bufo americanus American toad	Mid–April–early July in central Maine	/40°F	Long musical trill, 6–30 sec.
Rana palustris Pickerel frog	Late April or early May–May in central Maine	50–65°F /50–65°F	Low-pitched croak, somewhat like a rusty door opened slowly, or snoring grunt, similar to leopard frog but shorter and softer
Bufo woodhousii fowleri Fowler's toad *Bufo woodhousii woodhousii* Woodhouse's toad	After *B. americanus* B.w.f., April in N. Carolina, B.w.w., February in Arizona	65°F /65°F	Nasal "waaaaah" 1–5 sec. long, repeated 4–8 times per minute
Hyla versicolor Gray tree frog	May–August in central Maine	Above 68°F/	Short, very loud trill, lasting 0.5–3 seconds, repeated about 10 times in 30 sec.
Rana clamitans Green frog	May–August in central Maine	/70°F	Single explosive note like the plucking of a loose banjo string, may repeat 3–4 times, each note softer than the last
Rana catesbeiana Bullfrog	Last to breed in Maine, mid- to late June and July	68–72 °F/66 °F	Deep-throated "jug-o-rum"

Source: From information in Hunter, Albright, and Arbuckle, 1992; J. Markowsky, pers. comm.; Rugh, 1948, B.K. Sullivan, pers. comm.; and Wright and Wright, 1949.

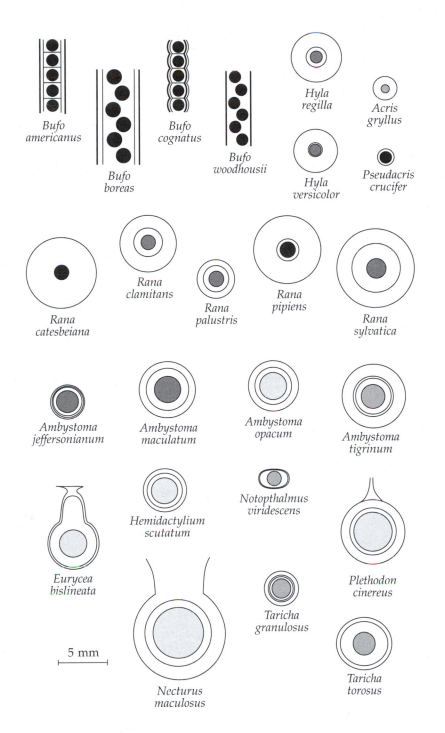

Figure 14.2
Eggs from some common North American amphibians, showing relative size and the diameter and number of jelly capsules. The central dark structure in each diagram is the egg, surrounded by its jelly capsules. The vitelline membrane, which lies snugly around the egg, is not shown. The relative darkness of the egg represents the amount of pigmentation that is found in the animal hemisphere of the egg. The most lightly shaded eggs are unpigmented. Those more darkly shaded are light-to-dark brown, and the darkest are black. The first two rows of eggs are those of common anurans. The lower rows are eggs of common urodeles. (After Bishop, 1941; Stebbins, 1951; Wright and Wright, 1949.)

Table 14.3 Stages in amphibian development

Stage	Description	*Rana sylvaticus* in hours at 18°C	*Rana pipiens* in hours at 18°C	*Ambystoma punctatum* in hours at 18°C
1	Unfertilized	0	0	0
2	Gray crescent	1	1	
3	2-Cell	2.5	3.5	15.5 hrs
4	4-Cell	3+	4.5	21 hrs
5	8-Cell	4.5	5.7	25 hrs
6	16-Cell	5+	6.5	27 hrs
7	32-Cell	6	7.5	31 hrs
8	Mid-cleavage	12	16	2 days
9	Late cleavage	16	21	
10	Early gastrula, dorsal lip forming	19	26	4 days
11	Mid-gastrula	24	34	4.5 days
12	Late gastrula	28	42	5 days
13	Neural plate	36	50	6 days
14	Neural folds	40	62	7 days
15	Rotation	45	67	
16	Neural tube	50	72	
17	Tailbud	58	84	
18	Muscular response	65	96	
19	Heartbeat	75	118	
20	Gill circulation, hatching	90	140	

Source: After Rugh, 1948.

Determine what stages of development your embryos are in, using Table 14.3 and Figures 14.3 and 14.4. If you were doing this for publication, you would use a staging series that is more specific to the group being studied. For our purposes, this simplified series can be used for both anuran and urodele embryos, though it was constructed by Roberts Rugh for the anurans *Rana pipiens* and *Rana sylvaticus*.

You will be lucky if you have stages as young as cleavage, for this will mean that, unless the water from which you collected the eggs was exceptionally cold (which slows down development considerably), the eggs probably were laid that day or the night before. If you really have hit the jackpot, you will have eggs that have just been fertilized. In these you will see a lightly pigmented gray crescent-shaped region, marking an area directly opposite the point of sperm entry. This **gray crescent** material is the precursor to the **notochordal** mesoderm. The first cleavage usually, but not always, bisects this gray crescent. The first, second, and fourth cleavage furrows start at the animal pole and work their way toward the vegetal pole. The third and fifth cleavage furrows are perpendicular to these. If you have mid-cleavage stages, notice that cells in the vegetal half are larger and less numerous than in the animal half. This is a consequence of yolk being concentrated in the vegetal half. As in any embryo, a large amount of yolk is an im-

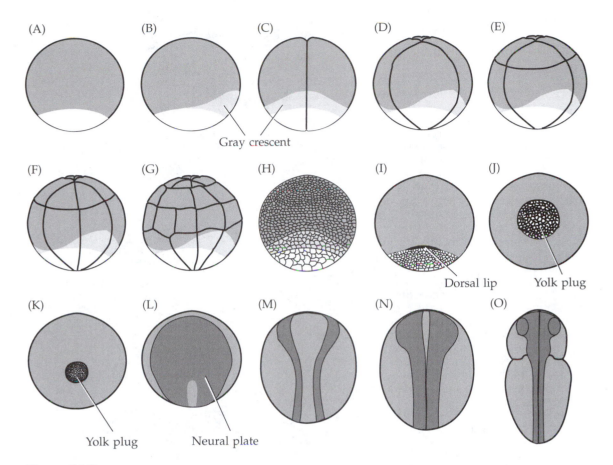

(A) (B) (C) (D) (E)

Gray crescent

(F) (G) (H) (I) (J)

Dorsal lip Yolk plug

(K) (L) (M) (N) (O)

Yolk plug Neural plate

Figure 14.3
Developmental stages of amphibians. Though there are a number of different staging series available for different species of anurans and urodeles, this simplified series (based on Rugh, 1948) can be used, with some exceptions, for most species of anurans and urodeles up to hatching. The diagrams illustrate the stages listed in Table 14.3. (A) Stage 1: Unfertilized egg. (B) Stage 2: Fertilized egg. There is a shifting of cytoplasm that uncovers a lighter area of pigmentation called the gray crescent. This is the precursor to the notochordal mesoderm. (Does not appear in the South African clawed frog, *Xenopus laevis*.) (C) Stage 3: 2 Cells. The first cleavage plane usually bisects the gray crescent. (D) Stage 4: 4 Cells. The second cleavage plane is parallel to the first. (E) Stage 5: 8 Cells. The third cleavage plane is perpendicular to the first two. (F) Stage 6: 16 Cells. The fourth cleavage plane is parallel to the first two. (G) Stage 7: 32 Cells. The fifth cleavage planes are parallel to the third. (H) Stage 8: Mid-cleavage. Cleavage is faster in the animal hemisphere than in the vegetal hemisphere, so the cells in the animal half are smaller and more numerous than those in the vegetal half. (I) Stage 10: Dorsal lip forms. Gastrulation is just beginning as cells involute forming the dorsal lip of the blastopore. (J) Stage 11: Mid-gastrula. The embryo is shown from the lower side. Cells are involuting from all sides of the blastopore. Large, yolky endoderm cells caught in the middle form the yolk plug. (K) Stage 12: Late gastrula. The embryo is shown from the lower side. As gastrulation proceeds, the yolk plug gets smaller. (L) Stage 13: Neural plate forms. The embryo is shown from the dorsal side. Neural ectoderm begins the process of forming a neural tube. (M) Stage 14: Neural folds form. The embryo is shown from the dorsal side. The edges of the neural plate are folding upwards. (N) Stage 15: Rotation begins. The embryo shown from the dorsal side. The neural folds are closing to form a neural tube. The epidermis of the embryo is ciliated, and the beating of the cilia cause the embryo to rotate within its vitelline envelope. (O) Stage 16: Neural tube. The embryo is shown from the dorsal side. The neural tube is now closed. The bulge of the forming eyes can be seen at the anterior end.

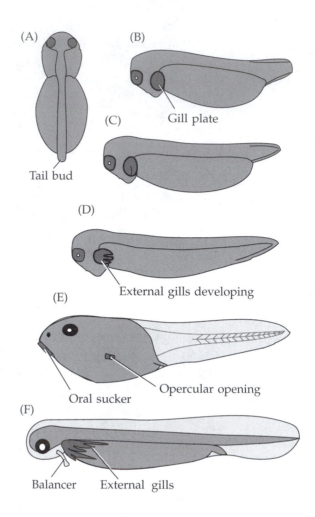

Figure 14.4
Stages in amphibian development continued. (A) Stage 17: Tail bud forms. The embryo is shown from the dorsal side. The tail begins forming. A short stubby tail bud is seen at the posterior end. (B) Stage 18: Muscular response begins. The embryo is shown in lateral view. The embryo and tail bud are elongating, and the gill plate, which will form external gills, is visible. Muscular movements have started, and the embryo can be seen to twitch within its vitelline envelope. (C) Stage 19: Heartbeat begins. The embryo continues to elongate, the gill plate is subdivided, and the heart has begun to beat. (D) Stage 20: Gill circulation and hatching. The embryo hatches from its vitelline envelope, the gill plate is further subdivided, and gill circulation is evident. (E) A generalized diagram of a young tadpole, the larva of an anuran. They possess oral suckers that allow them to attach. The gills are no longer visible, being enclosed in an opercular chamber with usually only one opening to the outside. (F) A generalized diagram of a young larva of a urodele. Called simply a salamander larva, it has a pair of balancers and visible external gills, which become long and featherlike as the larva grows. (After Rugh, 1948.)

pediment to cleavage. In the amphibian egg, this yolk slows down cleavage in the vegetal region, and by mid-cleavage the animal half has pulled ahead in its cleavage rate and ends up with more but smaller cells.

If your embryos have passed the cleavage stages, they may be in **gastrulation**. Look for the presence of a **blastopore**. In the amphibian, surface cells move toward the blastopore and move inward first at the **dorsal lip**, then the **lateral lips**, and finally at the **ventral lip** of the blastopore, a process called **involution**. Among amphibians, there is some variation in which cells move over the lips of the blastopore. In the South African clawed frog, *Xenopus laevis*, used in so many developmental studies, it is the endoderm; but in most North American amphibians, it is the mesoderm. In these, the dorsal lip is where the gray crescent material, or **notochordal mesoderm**, involutes. Once the entire blastopore is formed, a small plug of yolky endodermal cells is caught between the lips of the blastopore. This **yolk plug** is eventually internalized.

Following the major gastrulation movements, the **neural tube** starts forming from the **neural plate**. Look on the dorsal side of your eggs for evidence of an **invaginating neural plate**. The sides of the plate gradually meet in the midline and fuse to form the neural tube.

If your embryos are **rotating** within their vitelline envelope, then you know that they are already ciliated. This will occur just after the two sides of the neural plate have met in the midline. At this point, the brain should begin to look subdivided, and the bulges of the **optic cups** are obvious. Following this, the tail, starting as a short **tail bud**, begins to elongate as the entire embryo **elongates**. A **gill plate** will form, which eventually will be molded into **external gills**. Both anuran and urodele larvae form external gills, but the anurans secondarily enclose them in an **opercular chamber**, closing them off except for a single (though sometimes two) **opercular opening(s)**.

If your embryos are ready to hatch, they will be secreting a **hatching enzyme** that degrades the vitelline envelope and allows them to escape. You will know that hatching is near if the embryos are **contracting their musculature**, which will look like sudden twitches, and if their **heart has started beating**. If you have embryos this advanced, be sure to watch them carefully. Hatching is an exciting event to be witness to.

Urodele and anuran larvae initially look quite similar, but this similarity soon wanes as they continue their development. Urodele larvae, called simply salamander larvae, continue to elongate, their **gills become long and featherlike**, and a pair of **balancers** develops. These balancers aid in keeping the larva from sinking into the mud. Salamander larvae are carnivores and only eat food that is alive and wiggling. Hand-feeding salamander larvae becomes a game of making raw liver look like a wriggling insect larva by waving it on the end of a stick. Anuran larvae are called tadpoles. The body becomes ovoid followed by a long narrow tail. The intestines are especially noticeable, being long and coiled within the short body, making it look (I've always thought) like an hors d'oeuvre ready for a cracker. Tadpoles are vegetarians and need this large intestinal surface area to absorb nutrients. Tadpoles lack balancers but have an **oral sucker** that allows them to attach to sticks and other objects.

Environmental effects on amphibian development

Construct a chart showing the species of eggs that you found, percent survival, their developmental stage, any developmental defects observed, air temperature, and the temperature, pH, and condition of drying of their habitat. Look for specific effects due to acidity (described below), and add these to the chart.

One of the major effects of acidity on amphibian development is the loss of perivitelline space and the failure of the vitelline envelope to expand as the embryo elongates. The result is that the embryo becomes tightly coiled within the rigid, spherical envelope. In many cases, these coiled embryos never hatch. The acidity appears to inhibit the hatching enzyme. If they do manage to hatch, their vertebral columns are severely deformed and their tails are curled and bent. Other effects that have been noted are abnormal gill development and abnormal swelling near the heart. Physiologically, acidity disrupts the sodium and chloride balance within the embryo by inhibiting active uptake of these ions. Passive loss of the ions results in a net loss that eventually causes death.

Acidic conditions may also arrest development during early cleavage. In this case, the egg capsules become opaque and shrink. It has been reported that low pH can even prevent the first step in development—fertilization—by inactivating sperm, though the pH at which this occurs is debated. A pH lower than 4.0, however, is undoubtedly harmful to amphibian sperm.

Determine whether any of the abnormalities you found in your samples appear to be related to acid stress, and compare your results with those listed in Table 14.4. The table shows clearly that the range of pH tolerance varies among different species. What the table does not show is that this tolerance can vary even among different populations of a single species. For example, though the critical pH range for *Ambystoma maculatum* is given as 5–7 based on studies of several populations, another population was reported to have a critical range that started as low as pH 4.2. It appears that different populations are adapted to the natural pH of their habitat. For example, bogs typically are very acidic, with a pH as low as 4.2, so it can be expected that populations that lay their eggs in bogs will be more acid-tolerant than populations found in ponds or vernal pools. The data for

Table 14.4 pH tolerance of some common North American amphibians during development

Species	Lethal pH	Critical pH (<50% survival)	Optimal pH (maximal survival)
Ambystoma jeffersonianum embryo	4.0–4.6	4.0–5.0	5–6
Ambystoma maculatum embryo	4.0–5.0[a]	5.0–7.0[b]	7–9
Acris gryllus embryo	4.1	4.2–4.6	?
*Bufo americanus** embryo	3.5–3.9	4.0–4.5	?
*Hyla versicolor** embryo	3.8	3.9–4.3	?
*Pseudacris crucifer** embryo	3.8	4.0–4.2	?
*Rana catesbeiana** embryo	3.9	4.1–4.3	?
*Rana clamitans** embryo	3.8	3.8–4.1	?
*Rana palustris** embryo	4.0	4.2–4.4	?
Rana pipiens Sperm Embryo	4.5[c] 5.0	5.5[c] 6.0	6.5–7.0 6.3–7.5
Rana sylvatica Embryo Larva Metamorphosis	3.0–3.5[d] 3.25 3.25	3.75 3.5 (60%)[e] 3.5 (90%)	4.0–7.6 4–7.6 3.75–7.6

*Data from species living in a bog
[a]Failure to retract yolk plug, posterior trunk deformities, perivitelline space shrinks forcing embryo into tight coils.
[b]Swelling of chest near heart and abnormal gills.
[c]Measured in buffered solutions.
[d]Arrested in early cleavage, egg capsules become opaque and shrink.
[e]Abnormal, curled and bent tails.
Source: Based on data in Pierce, 1985; Pierce, et al., 1984; Pough, 1976; and Schlichter, 1981.

a number of the species in the table below came from populations laying in bogs. The same species that you find laying in non-bog habitats may be far less acid-tolerant than these. Can you determine probable lethal, critical, or optimal pH levels for your embryos from your data? If so, add these to Table 14.4, recognizing that this is only preliminary data.

Draw conclusions from your studies, and share these with the class. Data compiled from the entire class can be entered into your laboratory notebook. These data are valuable. Compiled over several years they may show trends that will be useful to environmental study groups. I invite you to send your carefully (of course) compiled results to an amphibian protection group in your area. The name of these varies among states and provinces, but you most likely will be able to find the name of one by calling your state or provincial Nongame and/or Endangered Species Program. If none is listed, look in the phone book for government listings to find numbers for a Department of Wildlife, Natural Resources, or Conservation that you can call instead.

Adopt an egg mass

Your estimates of developmental success will be greatly improved if you are able to return to a collecting site over a period of time and repeat your observations of percentage of survival within an egg mass, repeating temperature and pH readings and recording amount of drying of the habitat. To do this, it would be best to adopt one egg mass for continued monitoring. If time allows, you could test effects of ultraviolet radiation in the field by protecting some of the eggs using UV-blocking filters as discussed in Chapter 7. As a control, you can keep some of the eggs from the mass in the laboratory. Keep these eggs at a favorable temperature (18°C is usually adequate) and grow the embryos in Holtfreter's solution, a balanced salt solution designed especially for amphibian embryos. Compare the success rate of development between the embryos grown in the laboratory and those left in the field. Add these data to your chart. And remember at all times not to contribute to the problem of declining amphibian populations—never take large numbers of eggs from the field, do not expose embryos in the field to additional hazards, and return your collected embryos to their original habitats at the end of your studies.

Holtfreter's solution

NaCl	0.35 gm
KCl	0.005 gm
$CaCl_2$	0.01 gm
$NaHCO_3$	0.02 gm
Distilled water	100 ml

Stay in touch

For those who have now developed or have always had a love for these distant relatives of ours, you can keep abreast of the issues through a number of websites. One site maintained by the IUCN/SSC Declining Amphibian Populations Task Force, www.npwrc.usgs.gov/narcam/info/news/froglog2.htm, publishes the "Froglog Newsletter." Several others include Herp-Net, www.herpetology.com/herpnet.html, which is set up for international communication on amphibians and reptiles, www.frogweb.gov, a site for linking the public, press, and researchers, and www.im.nbs.gov/amphibs.html, a site maintained by the North American Amphibian Monitoring Program, which is looking for volunteers for frog calling surveys, terrestrial salamander monitoring, and creating amphibian atlases to document habitat.

Accompanying Materials

Tyler, M. S. and R. N. Kozlowski. 2000. *Vade Mecum: An Interactive Guide to Developmental Biology*. Sinauer Associates, Sunderland, MA. "Amphibian." This chapter of the CD shows in movies and labeled photographs the gametes, development, and metamorphosis of amphibians. Time-lapse photography of *Xenopus laevis* development is also included, as well as a gallery of adult frogs along with the calls of the males during mating season.

Gilbert, S. F. 2000. *Developmental Biology*, 6th Ed. Sinauer Associates, Sunderland, MA. Chapters 2, 10, and 18. Diagrams and discussion are given of amphibian life cycle, cleavage, gastrulation, and metamorphosis.

Fink, R. (ed.). 1991. *A Dozen Eggs: Time-Lapse Microscopy of Normal Development*. Sinauer Associates, Sunderalnd, MA. Sequence 8. This shows gastrulation in the South African clawed frog, *Xenopus laevis*.

Selected Bibliography

Bantle, J. A., J. Dumont, N. R. Finch and G. Linder. 1991. *Atlas of Abnormalities: A Guide for the Conduct of FETAX*. Oklahoma State Publications, Stillwater, OK. The FETAX system was developed to standardize testing of toxicity levels of environmental hazards on amphibians. By using the laboratory raised *Xenopus* frogs rather than native species, it avoids putting even more pressure on these endangered species.

Bishop, S. C. 1941. *The Salamanders of New York.* New York State Museum Bulletin No. 324. University of the State of New York, Albany, NY. This is a very detailed account of salamanders that occur in many regions besides New York. Specifics on breeding, egg masses, and individual eggs make it invaluable for species identification of eggs.

Bishop, S. C. 1943. *Handbook of Salamanders: The Salamanders of the United States, of Canada, and of Lower California*. Comstock Publishing, Ithaca, NY. This is one of the classics that all herpetologists know well. More conversational than his earlier work, it is both entertaining and an excellent reference.

Blaustein, A. R., J. M. Kiesecker, D. P. Chivers, D. G. Hokit, A. Marco, L. K. Belden and A. Hatch. 1998. Effects of ultraviolet light on amphibians: Field experiments. *Amer. Zool.* 38: 799–812. This excellent review of the effects of UV radiation on amphibians also discusses the synergistic effects of ultraviolet radiation and low pH, which can greatly increase mortality.

Blaustein, A. R. and D. B. Wake. 1995. The puzzle of declining amphibian populations. *Sci. Amer.* 272: 52–57. A beautifully illustrated article on a number of the threats to amphibian species, including UV radiation, destruction of habitat, pollution, disease, and human uses.

Doyle, R. 1998. Amphibians at risk. *Sci. Amer.* 279: 27. This short article shows maps of areas in the United States where amphibians are at risk.

Duellman, W. E. and L. Trueb. 1994. *Biology of Amphibians*. Johns Hopkins University Press, Baltimore. A general textbook about amphibians, covering anatomy, physiology, reproduction, ecology, and evolution. Its discussion of mating behaviors and breeding patterns is especially useful for this laboratory.

Elliott, L. 1998. *A Guide to Night Sounds* NatureSound Studio, P.O. Box 84, Ithaca, NY 14851-0084. This extremely useful collection is available as an audiocassette or CD and comes with a guide. It is a wonderful introduction for the novice. And it's cheap. It includes calls of ten of the most common anurans as well as night-calling birds, a few insects, and some mammals. It can be ordered from the NatureSound website, www.naturesound.com, or from North Sound by calling 1-800-336-6398.

Elliott, L. 1998. *The Calls of Frogs and Toads.* NatureSound Studio, P.O. Box 84, Ithaca, NY 14851-0084. More comprehensive than the selection above, this collection is available in both audiocassette and CD form. It e includes the sounds of 42 species of anurans along with a booklet on identification. It can be ordered from the NatureSound website, www.naturesound.com, or from North Sound by calling 1-800-336-6398.

Freda, J. 1991. Effects of acidification on amphibians. In *Acidic Deposition: State of Science and Technology, Vol. II. Aquatic Processes and Effects.* U.S. National Acid Precipitation Assessment Program, Washington, D.C., pp. 13-135–13-151. A good summary of the effects of acid rain on amphibian development, this both reviews and critiques early and more recent studies.

"Froglog." Newsletter from the World Conservation Union (IUCN), Species Survival Commision (SSC), Declining Amphibian Populations Task Force (DAPTF), and Center for Analysis of Environmental Change (CAEC). This is an extremely informative and concise publication available on the web at www.npwrc.usgs.gov/narcam/info/news/froglog2.htm.

Heyer, W. R., M. A. Donnelly, R. W. McDiarmid, L. C. Hayek and M. S. Foster (eds.). 1993. *Measuring and Monitoring Biological Diversity: Standard Methods for Amphibians.* Smithsonian Institution Press, Dept. 900, Blue Ridge Summit, PA 17294-0900, 1- 800-782-4612. Very recent, and inexpensive in its paperback version, this is an important book for anyone who would like to help in conducting amphibian counts for their geographic area and protecting local populations. It gives standardized methods for sampling and information on analyzing and using data.

Houck, L. D. 1998. Integrative studies of amphibians: from molecules to mating. *Amer. Zool.* 38: 108–117. This is a nice discussion of courtship pheromones used by male salamanders in courtship.

Howells, G., 1990. *Acid Rain and Acid Waters.* Ellis Horwood Limited, West Sussex, England. This has only a short section on the effects of acid rain on amphibians, but helps to give a general understanding of the global effects of acid rain.

Hunter, M. L., Jr., J. Albright and J. Arbuckle (eds.). 1992. *The Amphibians and Reptiles of Maine.* Maine Agricultural Experiment Station, University of Maine Press, Orono, ME.

Hunter, M. L., A. J. K. Calhoun and M. McCollough. 1999. *Maine Amphibians and Reptiles.* University of Maine Press, Orono, ME. All herp-enthusiasts, whether seasoned pros or the newly initiated, should have this book. It is an extremely complete compilation of information about amphibians and reptiles in Maine, but most of the information is applicable to many regions. It is delightfully written and beautifully illustrated with color photographs, pen-and-ink drawings, and includes a CD on calls of male frogs.

La Clair, J. J., J. A. Bantle and J. Dumont. 1998. Photoproducts and metabolites of a common insect growth regulator produce developmental deformities in *Xenopus*. *Environ. Sci. Technol.* 32: 1453–1461. A paper that uses the frog embryo teratogenesis assay-*Xenopus* (FETAX) method to determine the level of toxicity of a commonly used insecticide.

Pierce, B. A. 1985. Acid tolerance in amphibians. *BioScience* 35: 239–243. This is a very useful, easily understood review article, including a table and a long reference list.

Pierce, B. A., J. B. Hoskins and E. Epstein. 1984. Acid tolerance in Connecticut wood frogs (*Rana sylvatica*). *J. Herpetology* 18: 159–167. An important study on the effects of acid rain, it includes information on different stages of development as well as larvae.

Phillips, K. 1995. *Tracking the Vanishing Frogs: An Ecological Mystery*. Viking Penguin, New York. Geared toward a lay audience, this well-written account reveals the devastation to amphibian populations caused by human activities.

Pough, F. H. 1976. Acid precipitation and embryonic mortality of spotted salamanders, *Ambystoma maculatum*. *Science* 192: 68–72. One of the early articles on the effects of acid rain on amphibians, it is cited in all review articles on the subject.

Rugh, R. 1948. *Experimental Embryology: A Manual of Techniques and Procedures*. Burgess Publishing, Minneapolis, MN. An extremely valuable, out-of-print manual that was written by one of the venerable experts in amphibian development. The careful descriptions of techniques and developmental stages, illustrated with drawings and photographs, are invaluable to any embryologist. Look for it on out-of-print book lists, and snatch it up when you see it.

Schlichter, L. C. 1981. Low pH affects the fertilization and development of *Rana pipiens* eggs. *Can. J. Zool.* 59: 1693–1699. This study, though somewhat confusing, is the first to attempt to determine effects of acidity on amphibian sperm. The conclusions are now in some dispute, since buffered solutions were used and the buffers themselves might have been toxic to the sperm.

Stebbins, R. C. 1951. *Amphibians of Western North America*. University of California Press, Berkeley. An extremely thorough guide that includes information on breeding, egg masses, and larvae for each of the species covered.

Stebbins, R. C. 1985. *A Field Guide to Western Reptiles and Amphibians*, 2nd Ed. Peterson Guides, Princeton, NJ. This is an excellent guide, not as detailed as the volume above, but very handy in the field.

Tome, M. A. and F. H. Pough. 1982. Responses of amphibians to acid precipitation. In *Acid Rain/Fisheries. Proceedings of an International Symposium on Acid Precipitation and Fisheries Impacts in Northeastern North America, Ithaca, New York, August 2–5, 1981*. American Fisheries Society, Bethesda, MD, pp. 245–254. An older review article than that by Freda, this is still useful, including a number of charts and tables.

Twitty, V. C. 1966. *Of Scientists and Salamanders*. W. H. Freeman, San Francisco. A truly delightful account of one embryologist and his work, this stands as a valuable historical work and entertaining bedtime reading.

Tyning, T. F. 1990. *A Guide to Amphibians and Reptiles*. Little, Brown, Boston. This well-written, informative guide concentrates on relatively few species, but gives extremely useful information about their behavior and mating.

Wright, A. H. and A. A. Wright. 1949. *Handbook of Frogs and Toads of the United States and Canada*, 3rd Ed. Comstock Publishing, Ithaca, NY. What the Bishop handbook is for salamanders, this classic is for anurans. Illustrated with photographs of adults, larvae, and egg masses, it an important reference to add to any herpetologist's or naturalist's library.

Wyman, R. L. 1990. What's happening to the amphibians? *Cons. Biol.* 4: 350–352. This is an excellent summary of the problem of declining amphibian populations.

Suppliers

Cole-Parmer Instrument Company

625 East Bunker Ct.
Vernon Hills, IL 60061
1-800-323-4340
www.cole-parmer.com

Cole-Parmer has a large array of inexpensive to top-of-the-line pH meters. They also have inexpensive glass pocket thermometers.

Fisher Scientific

711 Forbes Ave.
Pittsburgh, PA 15219-4785
1-800-766-7000
www.fishersci.com

Fisher offers a large array of pH measuring devices. Among these are the most expensive and reliable Orion meters with Orion Ross electrodes, inexpensive and expensive pocket pH meters, colorimetric pH measuring kits, pH paper, and pH paper for soil testing. They also have a wide range of metal and glass pocket and field thermometers. You also can get 0.01 N HCl from Fisher and dilute it to 0.001 N (pH 4.0) for testing electrodes on low ionic acidic solutions.

Ward's Natural Science Establishment Inc.

P.O. Box 92912
5100 West Henrietta Road
Rochester, NY 14692-2660
1-800-962-2660
www.wardsci.com

Ward's offers a colorimetric pH measuring kit (for acid rain), an inexpensive portable water testing instrument that tests pH and temperature called Water Test Laboratory, and an acid rain test strip kit. They also have metal and glass pocket thermometers, from inexpensive to expensive.

Index